Microbial Transformations of Bioactive Compounds

Volume I

Editor

John P. Rosazza, Ph.D.
Professor and Head
Division of Medicinal Chemistry and Natural Products
College of Pharmacy
University of Iowa
Iowa City, Iowa

CRC Press, Inc.
Boca Raton, Florida

Library of Congress Cataloging in Publication Data
Main entry under title:

Microbial transformations of bioactive
compounds.

Includes bibliographical references and
indexes.
1. Microbiological synthesis.
2. Xenobiotic metabolism. I. Rosazza, John P.
QR88.3.M5 576'.11925 81-12281
ISBN 0-8493-6065-X (v. 1) AACR2
ISBN 0-8493-6066-8 (v. 2)

Direct all inquiries to CRC Press, Inc., 2000 Corporate Blvd., N.W., Boca Raton, Florida, 33431.

International Standard Book Number 0-8493-6065-X
International Standard Book Number 0-8493-6066-8

Library of Congress Card Number 81-12281
Printed in the United States

PREFACE

The microbial world consists of an incredibly diverse variety of living creatures which thrive by their remarkable abilities to adapt to an ever-changing environment. Part of this ability to adapt is attributable to the formation of enzymes as they are needed, and to the presence of certain constituitive enzymes. In general, these enzymes perform critical catabolic and anabolic functions for the organisms which produce them. Some of these functions are highly specific, and the enzymes involved are capable of interacting only with specific classes of substrates. Other enzymes possess relatively broad substrate specificities and these are commonly associated with biodegradative processes. The field which we know as microbial transformations has been largely concerned with attempting to understand and harness the enzymatic potential of microorganisms and to elucidate pathways and mechanisms operative in microbial transformation processes.

Biotechnological developments have occurred with staggering rapidity during the past decade and these have included advances in the microbial transformation field. Activity in the area of microbial transformations most probably reflects the availability of new analytical, chemical, and biochemical tools as well as our clearer perceptions of the nature of enzymes involved in metabolic interconversions. This is evident, for example, in the now widely-recognized presence of the cytochrome P-450 monooxygenase systems in all classes of microorganisms. Our present day technologies have enabled a more precise determination of metabolic pathways and of the mechanisms of enzymatic transformations of even the most complicated types of organic and inorganic substrates.

Many of the philosophies adapted to microbial transformation research were derived from earlier work with the steroids. Perhaps the greatest burst of research activity in the microbial transformation field began in the 1940s and extended over a period of three decades during which steroid transformations were vigorously and successfully developed. Processes were established on an industrial scale particularly with the steroid hormones whose combined microbiological and chemical syntheses are among the finest scientific endeavors of this century. Certainly the successes realized with the steroids were instrumental in stimulating microbial transformation research activity with other classes of physiologically active compounds like the alkaloids, antibiotics, cannabinoids, and the prostaglandins.

The tools and philosophies of microbial transformations have also been extended to research broadly concerned with problems of pharmacological and toxicological significance. Evidence for this may be seen in the extensive literature on biotransformations of environmentally important compounds like pesticides, aromatic hydrocarbons, and other xenobiotics. The utilization of microbial systems as tools in drug development work has been widely described under the "microbial models for mammalian drug metabolism" concept, and advances have been made in determining how microorganisms in the intestines achieve xenobiotic transformations of possible consequence to the host.

These volumes were assembled with the intent of bringing together contemporary research efforts in the microbial transformation field with an additional focus on physiologically active agents. Primary emphasis has been devoted to work reported during the past 10 years, and chapters have been prepared by individuals actively engaged in microbial transformation work. Notably absent from these volumes is a chapter dealing with the steroids, a subject which has been extensively reviewed elsewhere. The book begins with notes on the historical development of the microbial transformation field, and a chapter dealing with fundamental methodologies is also included. It is intended to provide a comprehensive and well-referenced look at the microbial transformation field as it exists today and to illustrate some of the current and future directions possible with it. If these volumes provide the stimulation necessary to lure others into this fascinating area of research, we will feel that it has been a success.

John P. Rosazza

THE EDITOR

John P. Rosazza, Ph.D., is Professor and Head of the Division of Medicinal Chemistry and Natural Products of the College of Pharmacy at the University of Iowa. He joined the faculty there as an Assistant Professor in 1969 following a year of postdoctoral study at the University of Wisconsin in the area of Pharmaceutical Biochemistry, and was promoted to the rank of Associate Professor in 1972 and full Professor in 1977. He received the B.Sc. degree in Pharmacy from the University of Connecticut in 1962, the M. Sc. degree in 1966 and the Ph.D. degree in 1968 from the same institution. Graduate degrees were received with research specialties in fermentation chemistry and organic chemistry. The Ph.D. degree was completed under the auspices of a training grant from the NIH in the area of Natural Products.

Dr. Rosazza has given service to several organizations. He serves as a referee for the Journal of Natural Products; the Journal of Pharmaceutical Sciences; Bioorganic Chemistry; Applied and Environmental Microbiology; and Antimicrobial Agents and Chemotherapy. He is a member of the editorial advisory boards for the Journal of Natural Products, Applied and Environmental Microbiology and Antimicrobial Agents and Chemotherapy. Dr. Rosazza has organized symposia for the American Society of Microbiology, the American Chemical Society and the American Society of Pharmacognosy. He is a member of the Executive Committee and is Librarian of the Division of Microbial and Biochemical Technology of the ACS. He is President of the American Society of Pharmacognosy. In addition, he is a consultant for the National Cancer Institute and for several industrial firms, and has served as a member of the Bio-Organic Natural Products study section for review of NIH grants.

Dr. Rosazza has pursued strong interests in microbial chemistry, biochemistry and natural products chemistry through the development of a vigorous program of research in the area of microbial and enzymatic transformations of organic compounds. Major emphases in his work have been on: microbial transformations of naturally occurring antitumor compounds; the applications of microbial systems as models for mammalian xenobiotic metabolism; and the development of microbial and enzymatic systems for use in organic synthetic processes. This work has been strongly supported by grants and contracts from the National Cancer Institute and from industry. Along with his fellow students and coworkers, Rosazza has published his work in numerous journals having published some 60 manuscripts to date.

CONTRIBUTORS

Jean-Marc Bollag, Ph.D.
Professor of Soil Microbiology
Pennsylvania State University
University Park, Pennsylvania

Patrick J. Davis, Ph.D.
Assistant Professor
Division of Medicinal Chemistry and Pharmacognosy
College of Pharmacy
University of Texas
Austin, Texas

Peter Goldman, M.D.
Maxwell Finland Professor of Clinical Pharmacology
Harvard Medical School
Boston, Massachusetts

Charles T. Goodhue, Ph.D.
Senior Research Associate
Biosciences Division
Eastman Kodak Company
Rochester, New York

Ching T. Hou, Ph.D.
Research Associate
Corporate Sciences Research Laboratories
Exxon Research and Engineering Company
Linden, New Jersey

Vincent P. Marshall, Ph.D.
Senior Scientist
The Upjohn Company
Kalamazoo, Michigan

George P. Peruzzotti, Ph.D.
Senior Research Scientist
Miles Laboratories, Inc.
Elkhart, New Jersey

Larry W. Robertson, Ph.D.
Associate Professor
Division of Medicinal Chemistry and Pharmacognosy
College of Pharmacy
Ohio State University
Columbus, Ohio

John P. Rosazza, Ph.D.
Professor and Head
Division of Medicinal Chemistry and Natural Products
College of Pharmacy
University of Iowa
Iowa City, Iowa

Oldrich K. Sebek, Sc.D.
Senior Scientist
The Upjohn Company
Kalamazoo, Michigan

Robert V. Smith, Ph.D.
Professor and Director
Drug Dynamics Institute
College of Pharmacy
University of Texas at Austin
Austin, Texas

Paul F. Wiley, Ph.D.
Senior Scientist
The Upjohn Company
Kalamazoo, Michigan

TABLE OF CONTENTS

Volume I

Volume II

Chapter 1

NOTES ON THE HISTORICAL DEVELOPMENT OF MICROBIAL TRANSFORMATIONS

Oldrich K. Sebek

TABLE OF CONTENTS

I. INTRODUCTION

As discussed in this book, an impressive amount of information on microbial transformations has been published, especially in the last three decades. In view of their theoretical significance and the success with which some of these reactions have been applied to the solution of practical problems, it may be of interest to recall some of the earlier papers which laid the foundation for this development and to sketch the beginnings of this important segment of microbiological endeavor.

Throughout history, microorganisms have been of considerable social and economic importance to man. Some of the processes for which they have been used, such as the leavening of bread, making of beer and cheese, saccharifying grains, retting flax and hemp, or unhairing hides, are age-old. The knowledge of others, such as the production of certain alcohols, organic and amino acids, antibiotics, vitamins, alkaloids, and nucleotides, is of a much more recent date. In time, it has also become recognized that microorganisms can modify some of these products by simple and chemically well-defined reactions and that these reactions are catalyzed by enzymes.

Reports on such reactions began to appear in the literature during the second half of the 19th century. However, at that time the association of chemical changes, particularly those of oxidative nature with microbes as their causative agents, was not of intense interest to investigators. Such interest was overshadowed by the studies on anaerobic fermentations which were defined by Pasteur as life without air and which occupied a considerable part of his illustrious career between 1857 and 1876. Toward the end of that period, it was discovered that yeasts contained enzyme invertase which converted sucrose to glucose and fructose, and more importantly, that the enzyme retained this activity even when extracted from cells. Although it was suggested by the famous French physiologist Claude Bernard that yeasts might contain similar extractable catalysts which would convert sugar to alcohol and CO_2, Pasteur's view of fermentation as a cellular process prevailed until 1897. In that year, Buchner demonstrated that cell-free extracts prepared by grinding yeast cells with sand could carry out alcoholic fermentation reactions in the absence of living cells.[1] Thus Bernard's concept was upheld and Buchner's discovery ushered in the new scientific discipline of biochemistry and enzymology.

About that time, a number of additional simple chemical reactions catalyzed by microorganisms were described in the literature. By the turn of the century, the amount of published information and its nature inspired Plimmer in 1903 to organize the literature concerned with simple microbial transformations into several distinct fermentation type reactions. His book[2] described oxidation, reduction, hydrolysis, nitrification, denitrification, simple fermentations, and the formation of optically active products. It is the prototype of many others which have since been published on this subject.

II. EARLY DEVELOPMENTS OF MICROBIAL OXIDATIONS

Vinegar production is perhaps the oldest and best known example of microbial oxidation and may be used to illustrate some of the important developments in the field of microbial transformations. Its preparation and use as a condiment have been known since the dawn of recorded history. Some 2000 years B.C., the Babylonians referred to vinegar as the fruit of the palm tree since it was prepared from the sap of the tree and also from the pressed juice of ripe dates, honey, and beer. Vinegar of Edom derived its designation from the biblical name for Arab desert.[3] Since sour (leavened) bread was used as a starter for its preparation, Arab Jews were forbidden to consume it during Passover. It was important to them, nevertheless, since according to the Talmud, "a man without salt and vinegar is a lost man".[4] Vinegar was known to the Chinese and the Egyptians as well. The Assyrians treated chronic middle-ear diseases with it and Hippocrates used it on his patients in 400 B.C. It is said that vinegar provided refreshment for the Greeks and Romans when diluted with water and flavored with spices. According to the New Testament, vinegar was offered to Jesus on the cross in the last moments of his earthly life and is mentioned twice in the Old Testament.[5] Its present-day manufacture originated toward the end of the 14th century in France, especially in the area of Orléans and the surface method employed in its preparation now bears the name of this city in which it originated.

Lavoisier demonstrated in 1793 that oxygen was necessary for the conversion of alcohol to vinegar and Dobereiner showed in 1823 that oxygen was actually consumed in this process. The pellicle which formed on the surface of such fluids was named *Mycoderma aceti* by Persoon in 1822 but its function in the acidification process remained unclear. Liebig thought the mycoderma to be nonliving and porous and that it possessed catalytic properties like the moistened platinum black used by Davy in 1821. Kützing concluded in 1837 that it was of a living nature.

In 1862, Pasteur[6] investigated this pellicle which he called "the flower of vinegar" and concluded that it "serves as a transport of air oxygen to a multitude of organic substances". The subsequent work by Brown in 1886 confirmed Pasteur's findings, namely that the alcohol oxidation proceeded through acetaldehyde and that CO_2 and H_2O were the final products obtained when acidification progressed beyond the acid stage.[7] Brown named the causative agent *Bacterium xylinum* and also found that it could oxidize propanol to propionic acid, and mannitol to fructose. In addition, he confirmed earlier findings reported by Boutroux in 1878 and 1880 that glucose was converted into gluconic acid which was first thought to be lactic acid.[8,9]

Bacterial oxidations of this kind attracted the attention of Bertrand who had previously carried out studies on the oxidation of monophenols, aromatic amines, and tyrosine by enzymes isolated from plants.[10] A paper published by him in 1896 noted that juice of the mountain ash berries could be infected by bacteria which he believed to be very similar or identical to the *Bacterium xylinum* described by Brown. They were carried by fruit flies and converted sorbitol in mountain ash berry juice to sorbose.[11] This sugar was identical to the compound isolated by Pelouze in 1852 from the juice of same origin that had been allowed to stand in the air for more than 1 year.[12] In a rapid succession, Bertrand reported that eight other sugar alcohols could be oxidized to their corresponding keto sugars. Of these, the oxidations of mannitol to fructose, glycerol to dihydroxyacetone, erythritol to erythrulose, and perseitol to perseulose were more prominent.[13] Since the reaction mechanism in all of these cases appeared to be the same, Bertrand formulated a rule which bears his name today. The rule describes the structural characteristics of polyhydric alcohols required if they are to be metabolized by *B. xylinum*. With minor modifications, which accomodate the oxidation of cyclic alcohols, the rule is still valid today.[14,15]

Bacteria with these properties are readily isolated from the beer that had turned sour when exposed to the air for an extended length of time. Their isolation was one of the initial experiments for new students in Professor Kluyver's laboratories in Delft.[16] They were usually identified as members of *Acetobacter rancens* but de Leeuw, one of Kluyver's students, isolated in 1923 a different bacterium of this group which quite unexpectedly oxidized glucose to 5-ketogluconic acid. Although this bacterial reaction was not new (it had been discovered by Boutroux[17] in 1886), the causative organism was found to be a new bacterial species. To emphasize its ability to carry out limited oxidations, it was named *Acetobacter suboxydans*.[18] When L-sorbose assumed prominence in the middle 1930s as an intermediate in the Reichstein synthesis of L-ascorbic acid,[19] this bacterium was employed in the highly efficient preparation of this sugar from D-sorbitol. The economic advantage of this fermentation type-reaction was greatly advanced by the application of newly developing submerged growth methodology.[20,21] This technique was reported to be used by Takamine as early as the 1890s to hydrolyze starch to fermentable sugars[22] and by Calmette in 1902 to prepare gallic acid from tannin extracts.[23] Wehmer tried to produce citric acid by this method in 1912, but yields were lower than those obtained under stationary surface conditions.[24] Others were similarly unsuccessful, but the breakthrough occurred in 1928 to 1930 when two reports revealed that a strain of *Aspergillus fumaricus* produced 4 to 6 times more gluconic acid in agitated and aerated cultures than in nonaerated stationary surface controls.[25,26] These and other reports stimulated further investigations on the role of oxygen supply and its efficient transfer to cells in submerged nutrient media. These early developments were ultimately combined with strain improvement programs and fermentation optimization protocols, and laid the foundation for the large-scale antibiotic and amino acid fermentations and steroid transformations.

III. PURE CULTURE COLLECTIONS

Most of the early microbial transformation work was carried out with cultures isolated from natural sources. These cultures were poorly preserved and often lost. As a result, the work with such microorganisms could be neither repeated nor extended. This problem created the need for central repositories where cultures could be stored and maintained in their physiologically and genetically unchanged state and from where they would be available to other investigators. Guided by this concern, the first culture collection was established in Utrecht by the International Association of Botanists in 1906. It was later transferred to Delft and eventually to Baarn where it has served the scientific community ever since. Many other culture collections have subsequently been established and have thus assured the reproducibility of microbiological and biochemical research efforts to the highest degree.

The requirement for the pure culture was soon shown to be essential also for the successful large-scale fermentative production of acetone and butanol. While the manufacture of ethanol, lactic acid, and vinegar could be carried out with cultures that are not completely pure, the yields of acetone and butanol were drastically reduced when the producing organisms were contaminated by others, especially by lactic acid bacteria. In 1912, Chaim Weizmann discovered a superior culture *(Clostridium acetobutylicum)* for the production of these alcohols. This fermentation, which was patented in 1915,[27] was rapidly developed into an efficient process and became an important source of acetone as solvent for nitrocellulose in the manufacture of smokeless powder during World War I. It is of considerable historical significance that through this outstanding contribution to the Allied war effort, Weizmann was instrumental in securing the Balfour Declaration in 1917. In this document, Great Britain pledged itself to support the establishment of a national home for the Jewish people in Palestine and resulted in the creation of the State of Israel in 1948 with Weizmann as its first president.

IV. OTHER MICROBIAL TYPE REACTIONS: REDUCTION, HYDROLYSIS, CONDENSATION

At the time of the publication of Plimmer's book[2] in 1903, only a few examples of microbial reduction were known: the formation of hydrogen sulfide from sulfur, thiosulfate and thioglycolate, and also from sulfate via sulfite in a two-step reaction sequence. Conversions of tartaric and malic acids to succinic acid by bacteria[28] and reductive methylation of arsenic in arsenic-containing paints by *Penicillium brevicaule* to the poisonous product later identified as trimethylarsine were also reported.[29]

Hydrolysis was another early reaction of importance. The retting of flax and hemp to free bast fibers used in making linen, has been practiced for thousands of years. It is carried out by butyric acid bacteria which develop when plant stems are kept immersed in water. The bacteria decompose plant pectins whereby the stem structure is loosened and the cellulosic fibers are freed. By a similar technique, diosgenin, a valuable raw material for the synthesis of steroid hormones, is produced today in Mexico from the roots of the barbasco plant *(Dioscorea).* The roots are chopped into small pieces and immersed in water in deep pits. By spontaneous fermentation and hydrolysis of glycosides, diosgenin is freed from the plant tissue and recovered for use in the synthesis of a variety of steroid hormones.

Microbial hydrolysis of starchy materials has been employed for ages also in the preparation of foods and beverages. Bacterial vibrios were used by the South American Indians to make chicha, a wine-like drink from corn.[30] *Aspergillus oryzae* was used for centuries in the Orient to prepare koji and "Chinese yeast" for wines and spirits in China and Indochina. Chinese yeast was a mixture of amylolytic and fermenting microorganisms. It was said that it could be purchased as little flat cakes the size of a "five-franc coin", had an odor of mildewed flour, and was spiced with cinnamon and other herbs.[31] The amylo process used fungal amylase instead of malt diastase to hydrolyze grain starches and was said to have been in operation since 1891. Takamine modified it in 1914 by growing *Aspergillus niger* on wheat bran instead of steamed rice which was used to prepare koji. He called this material Taka-koji, and his experiments with *A. niger* grown in a cylinder similar to a pneumatic malt drum are among the earliest examples of submerged aerated fermentations.[22]

The hydrolysis of starch did not always proceed to completion to yield monosaccharides. Thus in 1903, Schardinger observed that *Bacillus macerans* converted starch into the dextrins and compounds consisting of six or more glucose units.[32] Microbial hydrolysis of di- and trisaccharides, cellulose, glucosides, esters, and proteins was known at that time, and investigations on the hydrolysis of tannin were also carried out. In 1867, Van Tieghem was interested in an earlier work by Pelouze who noted that gallic acid was slowly formed when tannin transformations were carried out by *Penicillium glaucum* and also by an aspergillus. This unidentified fungus was investigated by Van Tieghem who later wrote that this organism "appeared to me and to Dr. Léveille, whose great authority I am happy to invoke here, to constitute a distinct and new species which we name *Aspergillus niger*."[33] This is probably the first reference to this fungus which has since been one of the most cited microorganisms in the microbiological literature. Numerous other hydrolyses have been since described. The most prominent among them is the cleavage of penicillin to 6-aminopenicillanic acid which in turn serves as a unique starting material for the preparation of therapeutically valuable semisynthetic penicillins and cephalosporins. The hydrolysis is efficient, superior to chemical methods, and has been the only practical way of preparing this crucial intermediate.

Following World War I, microorganisms were used primarily for basic studies on intermediary metabolism by a number of illustrious investigators including Harden, Fernbach, Neuberg, Nord, Meyerhof, Kluyver, van Niel, and others. In addition, new examples of simple and useful microbial reactions were also reported. One, which was discovered by Neuberg and Hirsch in 1921, involved the condensation of benzaldehyde with acetaldehyde by yeasts to form optically active L-phenyl-l-hydroxy-2-propanone.[34] This product was in turn converted chemically into D-(-)ephedrine and the sequence was developed into a process in which microbial and chemical steps were combined into a successful industrial operation.[35]

V. STEROID TRANSFORMATIONS

An incidental observation by Lintner and von Liebig in 1911 that yeasts reduced furfuraldehyde to the corresponding alcohol[36] led Neuberg to investigate microbial reductions of other kinds of organic compounds. This microbial methodology was applied to the preparation of adrenal hormones by Mamoli and Vercellone, who was one of Neuberg's students. These investigators demonstrated in 1937 that yeasts reduced the nuclear ketone of 4-androstenedione to the male hormone testosterone and that a few other androstane derivatives were reduced in the same way.[37] A similar reduction of 3,12-diketocholanic acid was reported by Kim at about the same time.[38] These findings and the conversion of cholesterol to 7-hydroxycholesterol published 10 years later by Kramli and Horvath[39] clearly demonstrated that steroids and sterols could be metabolized by microorganisms in spite of their inherently poor solubilities in water. Thus, the stage was set for the momentous observation reported by Peterson and Murray in 1952 that *Rhizopus arrhizus* and other fungi of the order Mucorales convert progesterone to 11α-hydroxyprogesterone.[40,41] This microbial hydroxylation simplified and drastically improved the efficiency of multi-step chemical syntheses of corticosteroid hormones and their derivatives. In addition to its considerable practical value, this reaction and the related 16α-hydroxylation of progesterone by an unidentified actinomycete described shortly thereafter by Perlman, Titus, and Fried,[42] were also of signal theoretical significance. They pointed out quite persuasively the advantages of microbial enzymes as catalysts in performing stereospecific reactions that were difficult, inefficient, and in some cases impossible to perform by strictly chemical means. They caused an unprecedented surge in the application of microbial enzymes in chemical syntheses.

In subsequent years, thousands of microorganisms were examined and some were used with remarkable specificities in transforming different classes or organic compounds and natural products. Such substrates included not only polyhydric alcohols, sugars, steroids and sterols, but also alkanes, olefins, alicyclics, aromatics, terpenoids, alkaloids, antibiotics, antitumor agents, and pesticides. The spectrum of the type-reactions obtained with enzymatic systems has now been widened to include not only oxidation, hydroxylation, reduction, hydrolysis, and condensation, but also others such as dehydrogenation, β-oxidation, amination, deamination, hydration, dehydration, *N*- and *O*- dealkylation, decarboxylation, acylation, glycosylation, and isomerization.

The success with which microorganisms carry out the desired reactions should stress the advantages of their application, especially in asymmetric syntheses, and in those instances where a high degree of reaction specificity is essential. Although they have been used in the past as mere catalysts of simple but specific reactions, they also represent an attractive material for future studies on regulation, control, and genetic manipulation of such transformations.

REFERENCES

1. **Buchner, E.,** Alkoholische Gahrung ohne Hefezellen, *Ber. Dtsch. Chem. Ges.,* 30, 117, 1110, 1897.
2. **Plimmer, R. H. A.,** *The Chemical Changes and Products Resulting from Fermentation,* Longmans, Green and Company, London, 1903.
3. **Huber, E.,** Der Essig in der altbabylonischen Kulturgeschichte, *Dtsch. Essigind.,* 31, 12, 28, 1927.
4. *The Jewish Encyclopedia,* Funk & Wagnalls Company, New York, 1906.

5. *The Holy Bible*, Revised Standard Version, Psalms 69:21, Proverbs 10:26, Matthew 27:48, Mark 15:36, Luke 23:36, John 19:29, Collins' Clear-Type Press, New York 1952.
6. **Pasteur, L.**, Études sur les mycodermes. Rôle de ces plantes dans la fermentation acétique, *Compt. Rend.*, 54, 265, 1862; Suite a une précédente communication sur les mycodermes; Nouveau procéde industriel de fabrication du vinaigre, *Compt. Rend.*, 55, 28, 1862.
7. **Brown, A. J.**, The chemical action of pure cultivations of *Bacterium aceti*, *J. Chem. Soc.*, 49, 172, 1886.
8. **Boutroux, L.**, Sur la fermentation lactique, *Compt. Rend.*, 86, 605, 1878.
9. **Boutroux, L.**, Sur une fermentation nouvelle du glucose, *Compt. Rend.*, 91, 236, 1880.
10. **Bertrand, G.**, Sur les rapports qui existent entre la constitution chimique des composés organiques et leur oxydabilité sous l'influence de la laccase, *Compt. Rend.*, 122, 1132, 1896; Sur une nouvelle oxydase, ou ferment soluble oxydant d'origine végétale, *Compt. Rend.*, 122, 1215, 1896.
11. **Bertrand, G.**, Préparation biochimique du sorbose, *Compt. Rend.*, 122, 900, 1896.
12. **Pelouze, J.**, Sur une nouvelle matière sucrée extraite des baies de sorbier, *Ann. Chim. Phys.*, 35, 222, 1852.
13. **Bertrand, G.**, Action de la bactérie du sorbose sur les alcools plurivalents, *Compt. Rend.*, 126, 762, 1898; Sur le produit d'oxydation de la glycérine par la bactérie du sorbose *Compt. Rend.*, 126, 842, 1898; Préparation biochimique de la dioxyacétone cristalliseé, *Compt. Rend.*, 126, 984, 1898; Sur l'oxydation de l'érythrite par la bactérie du sorbose; production d'un nouveau sucre: l'érythrulose, *Compt. Rend.*, 130, 1330, 1900; Un nouveau sucre cristallisé, le perséulose à sept atomes de carbone, *Compt. Rend.*, 147, 201, 1908.
14. **Hann, R. M., Tilden, E. B., and Hudson, C. S.**, The oxidation of sugar alcohols by *Acetobacter suboxydans*, *J. Am. Chem. Soc.*, 60, 1201, 1938.
15. **Magasanik, B., Franzl, R. E., and Chargaff, E.**, The stereochemical specificity of the oxidation of cyclitols by *Acetobacter suboxydans*, *J. Am. Chem. Soc.*, 74, 2618, 1952.
16. **Kamp, A. F., La Rivière, J. W. M., Verhoeven, W., Eds.**, *Albert Jan Kluyver, His Life and Work*, North-Holland, Amsterdam, 1959, 80.
17. **Boutroux, L.**, Sur une fermentation acide du glucose, *Compt. Rend.*, 102, 924, 1886; Sur l'acide oxygluconique, *Compt. Rend.*, 111, 185, 1890.
18. **Kluyver, A. J. and de Leeuw, F. J.**, *Acetobacter suboxydans*, een merkwaardige azijnbacterie, *Tijdschr. Verg. Geneesk.*, 10, 170, 1924.
19. **Reichstein, T. and Grüssner, H.**, Eine ergiebige Synthese der ℓ-Ascorbinsäure (C-Vitamin), *Helvetica Chim. Acta*, 17, 311, 1934.
20. **Wells, P. A., Stubbs, J. J., Lockwood, L. B., and Roe, E. T.**, Sorbose from sorbitol. Production by submerged growths, *Ind. Eng. Chem.*, 29, 1385, 1937.
21. **Wells, P. A., Lockwood, L. B., Stubbs, J. J., Roe, E. T., Porges, N., and Gastrock, E. A.**, Sorbose from sorbitol. Semiplant-scale production by *Acetobacter suboxydans*, *J. Am. Chem. Soc.*, 31, 1518, 1939.
22. **Takamine, J.**, Enzymes of *Aspergillus oryzae* and the application of its amyloclastic enzyme to the fermentation industry, *J. Ind. Eng. Chem.*, 6, 824, 1914.
23. **Calmette, A.**, German Patent 129, 164, 1902.
24. **Wehmer, C.**, Über Citronensäuregärung, *Chemiker-Zeitung*, 36, 1106, 1912.
25. **Schreyer, R.**, Sauerungsversuche mit dem Pilz *Aspergillus fumaricus*, *Biochem. Z.*, 202, 131, 1928.
26. **Thies, W.**, Untersuchungen uber den Einfluss der Bedingungen auf die Saurebilung des Schimmelpilzes *Aspergillus fumaricus*, *Zentralbl. Bakteriol. Parasitenk.*, 11, 82, 321, 1930.
27. **Weizmann, C.**, Acetone and butyl alcohol, British Patent 4,845, 1915.
28. **König. F.**, Ueber Gährung der Weinsäure, *Ber. Dtsch. Chem. Ges.*, 14, 211, 1881; Darstellung der Bernsteinsäure aus Weinsäure durch Gährung, *Ber. Dtsch. Chem. Ges.*, 15, 172, 1882.
29. **Challenger, F., Higginbottom, C., and Ellis, L.**, Formation of organometalloid compounds by microorganisms. I. Trimethylarsine and dimethylarsine, *J. Chem. Soc.*, 95, 1933
30. **Marcano, V.**, Fermentation de la fécule. Présence d'un vibrion dans la graine de mais qui germe et dans la tige de cette plante, *Compt. Rend.*, 95, 345, 1882; Fermentation directe de la fécule. Mécanisme de cette métamorphose, *Compt. Rend.*, 95, 856, 1882.
31. **Calmette, A.**, Contribution à letude des ferments de l'amidon — La levure chinoise, *Ann Inst. Pasteur*, 60, 604, 1892.
32. **Schardinger, F.**, Über thermophile Bakterien aus verschiedenen Speisen und Milch, sowie über einige Umsetzungsprodukte derselben in kohlenhydrathaltigen Nährlösungen, darunter krystallisierte Polysaccharide (Dextrine) aus Stärke, *Z. Untersuch. Nahrungs-Genussmittel*, 6, 865, 1903.
33. **Van Tieghem, P.**, Sur la fementation gallique, *Compt. Rend.*, 65, 1091, 1867.
34. **Neuberg, C. and Hirsch, J.**, Über ein Kohlenstoffketten knüpfendes Ferment (Carboligase), *Biochem. Z.*, 115, 282, 1921.

35. **Hildebrandt, G. and Klavehn, W.,** ℓ-1-Phenyl-2-methylamino-1-propanol. U.S. Patent 1, 956, 950, 1934.

36. **Lintner, C. J. and von Liebig, H. H.,** Über die Reduktion des Furfurols durch Hefe bei der alkoholischen Gärung, *Z. physiol. Chem.*, 72, 449, 1911.

37. **Mamoli, L. and Vercellone, A.,** Über die biochemische Hydrierung des Dehydro-androsteron, *Z. physiol. Chem.*, 245, 93, 1937 Biochemische Umwandlung von Δ^4-Androstendion in Δ^4-Testosteron. Ein Beitrag zur Genese des Keimdrüsenhormons, *Ber. Dtsch. Chem. Ges.* 70, 470, 1937; Biochemische Umwandlung von Δ^4-Androstendion in Isoandrostandiol und Δ^4-Testosteron. Weiterer Beitrag zur Genese der Keimdrüsenhormone, *Ber Dtsch. Chem. Ges.*, 70, 2079, 1937.

38. **Kim, C. H.,** Über die enzymatische Hydrierung der Dehydrodesoxycholsäure durch Hefe, *Enzymologia*, 4, 119, 1937 and 6, 108, 1939.

39. **Krâmli, A. and Horvâth, J.,** Microbiological oxidation of sterols, *Nature (London)*, 162, 619, 1948 and 163, 219, 1948.

40. **Peterson, D. H. and Murray, H. C.,** Microbiological oxygenation of steroids at carbon 11, *J. Am. Chem. Soc.*, 74, 1871, 1952.

41. **Murray, H. C. and Peterson, D. H.,** Oxygenation of steroids by Mucorales fungi, U.S. Patent 2, 602, 769, 1952.

42. **Perlman, D., Titus, E., and Fried, J.,** Microbiological hydroxylation of progesterone, *J. Am. Chem. Soc.*, 74, 2126, 1952.

Chapter 2

THE METHODOLOGY OF MICROBIAL TRANSFORMATION OF ORGANIC COMPOUNDS

Charles T. Goodhue

TABLE OF CONTENTS

I. INTRODUCTION

Microbial transformations (also called "biotransformations" and "bioconversions") are chemical reactions performed by microorganisms. The reactions themselves are catalyzed by the enzymes within (or that orginate within) the microbial cell. Most of these enzymes are necessary for the normal functioning of the life processes of metabolism and reproduction. Some bear no obvious relation to these functions. This may be owing to our ignorance, or perhaps the enzymes are vestigal without a function crucial to the life of the cell. The objective of the art and science of microbial transformation is to use the cellular enzyme equipment to effect useful chemical reactions. Most of the technology with which we are concerned is directed to the use of enzyme activities in intact cells. These cells may be live, actively growing, or in a resting state in which cell division occurs only to a small extent, if at all. Dried cells also can be used effectively whether or not they are viable. Spores can be used. The use of enzyme mixtures extracted from cells is discussed briefly in this chapter. Techniques for the purification of enzymes are left to the province of enzymology. Of the 3000 or more enzymes present in a cell, the practitioner of microbial transformations endeavors to use just one, or perhaps only a few to obtain the desired results. Rather remarkable chemical syntheses have been accomplished in this manner. This chapter will attempt to explain in a practical sense the methods that have evolved to perform chemical reactions with microorganisms.

II. TYPES OF MICROORGANISMS USED IN BIOCONVERSIONS

For the person not trained in microbiology who wishes to become familiar with the methods of bioconversions, it is well that he learns the rudiments of the biology of the organisms that are used. This is not as difficult as it might seem, for most of the microorganisms used over the years have been selected from just three major groups: the bacteria, the fungi, and the yeasts. For that matter, the yeasts are classed as fungi, so there are really only two groups. In order to understand these organisms better, one should know how they are classified in relation to the other organisms of the living world. Taxonomy is the science of classification and naming. It has evolved a hierarchal system of arranging all living things into naturally related groups based on structure, biochemistry, and genetics. The three largest divisions are those of the kingdoms of the plants, the animals, and the protists. In this chapter we are concerned mainly with the protists. Microbiology is the science of the protists. Although we recognize that useful chemical reactions and syntheses can be performed by tissue cultures of plant and animal cells (and that in this case we would be invading the other kingdoms of the plants and the animals), the extra difficulties of tissue culture are such that the beginner would be advised to start his bioconversion efforts with the protists. The reader, having learned to exploit the chemical wonders of the microorganisms, then may venture into the more difficult area of tissue cell culture.

A. Taxonomy

As stated earlier taxonomy is the science of naming and classification of creatures. The names of individual microorganisms, as well as those of the higher plants and animals, consists of two parts: the genus and the species. The species is the fundamental biological classification. Members of the same species show persistent, basic differences from members of allied species in the same genus. The genus name is capitalized; the species is not. Both are italicized. An example is *Saccharomyces cerevisiae,* a familiar member of the yeasts. Groupings of related genera are called families. Related families are orders; groups of orders are classes, and these are further organized hierarchically into divisions and kingdoms. In many instances, it is sufficient to use the binomial genus and species name for a particular microorganism, but there are times when the further use of a strain or variety designation is required to specify mutants or particular variants within a species that have special properties. An example is *Saccharomyces bisporus* var. *mellis.*

1. Fungi

The kingdom of the protists includes the algae, the protozoa, the fungi, and the bacteria. The protists are subdivided into eukaryotic and prokaryotic organisms. Very simply, the eukaryotic cells contain a defined nucleus; the prokaryotic cells do not. Other differences in the two groups will be mentioned as we proceed with our discussion. Although the eukaryotic cells are inherently more complicated then their smaller cousins, the fungal taxonomy in outline is relatively straightforward compared to bacterial taxonomy. Algae, protozoa, and yeasts are the classes of eukaryotic protists. In bioconversions the fungi and yeasts are by far the most important of these. The fungi (sometimes called "the higher fungi" to distinguish them from bacteria, "the lower fungi") are nonphotosynthetic, predominantly coenocytic cells that grow in filamentous, branched structures called mycelia. Coencytic cells are multinucleate. That is, during growth which proceeds from a single cell or spore containing a single nucleus, the cell mass increases without apparent cell division; only the nuclei divide. The cell mass of some fungi can reach macroscopic size, as exemplified by the common mushrooms. The important thing to remember is that sometime during its life cycle the organism exists as a single microscopic cell with a single nucleus. Mycelium is the branched, filamentous outgrowth from a spore. Vegetative outgrowth also can proceed from mycelial fragments. From the mass of vegetative mycelium, fruiting structures arise which produce spores. The life cycles of the various fungi can be rather complicated, though in broad outline they consist of the outgrowth from spores, maturation, and sporulation, with sometimes a sexual phase intervening. Descriptions of life cycles can be found in microbiological reference works.[1,2] It is recommended that the life cycles of the fungi to be used for experiments be known to the investigator.

There are four classes of fungi: *Zygomycetes (Phycomycetes), Ascomycetes, Basidiomycetes,* and the *Fungi Imperfecti* (sometimes called *Deuteromycetes). Zygomycetes* reproduce from small endogenous asexual spores contained usually in sporangia. They closely resemble some of the algae, but of course they are not photosynthetic. Downy mildews and some of the black molds are examples of this class. Another important feature is the lack of septal divisions in the mycelium. When sexual spores are produced, they are borne exposed. *Ascomycetes* have a septate mycelium. Sexual spores are formed in saclike structures (asci) and are discharged by rupture of the sac. Included in this class are the common bread molds, citrus molds, as well as some of the yeasts. Fungi in this class also reproduce by means of asexual spores. *Basidiomycetes* also are septate, but their fruiting body structures for sexual spore production are more complicated still. The spores are borne on the surface of club-like structures called basidia. They are held there on short, slender stalks. At maturation they are released explosively to the environment by violent rupture of the structure. These fungi include mushrooms, puffballs, smuts, and some yeasts as well. The basis for the distinction between an ascomycete and a basidiomycete is their method of sexual sporulation. This is very convenient, but unfortunately, not all fungi can be observed to undergo a sexual stage. Fungi that have not been observed in a sexual stage are assigned to a catch-all category, the *Fungi Imperfecti.* If a sexual stage is ever observed, these fungi are properly classified.

2. Yeasts

Important members of the fungi are the yeasts. While there are relatively few species, yeasts play a very important part in commerce and industrial microbiology. The yeasts are predominantly unicellular, and they reproduce by budding, except for one at least, that reproduces by binary fission *(Schizosaccharomyces).* Some yeasts by conjugation of ascospores produce sexual spores. Representative species are found from the ascomycetes, the basidiomycetes, and the fungi imperfecti.

Eukaryotic cells, in addition to being distinguished by the possession of a nucleus, are much larger than prokaryotic cells. For example, in terms of cell volume yeasts are 20 to 50 μm^3 compared to the average cell volume of most true bacteria of 1 to 5 μm^3. Moreover, the cell or mycelium diameter of the higher fungi is approximately ten times that of the bacteria.

3. Bacteria

While the bacteria are smaller and somewhat simpler in design, their classification into the higher taxa of families and orders turns out to be difficult. This is mainly because there are so many species with different points of commonality that it is difficult to develop a consistent heirarchal system. The most ambitious attempt to do this is published as Bergey's Manual.[3] Correlations between major structural and functional characteristics are few and irregular. However, Bergey's has divided the bacteria into 5 orders, 42 families, and 225 genera.

The prokaryotic protists consist of the blue-green algae (cyanophytes) and the bacteria. Biotransformation procedures have employed mainly the bacteria, and so the blue-green algae will not be part of our discussion. The bacteria are a large and varied group which can be classified conveniently into three principal subgroups: the eubacteria, the myxobacteria, and the spirochetes. By far the most important group for bioconversions is the eubacteria, at least up to the present. This group, incidentally, is the largest and most varied of the three.

The following is a very simplified discussion of the types of the bacteria, without attempting to delve in depth into the complicated taxonomy. Since bacteria are difficult to classify, it is usually sufficient to recognize the bacterial genera and species. A good reference work for this is Skerman's Guide.[4] Bacteria are single-celled and microscopic in size. They reproduce by binary fission. Possibly the single most useful characteristic for subdivision of bacteria into groups is the Gram reaction. This reaction divides the bacteria into two large groups, Gram-positive and Gram-negative, based on dye reactions with the cell wall.

Another way of classifying bacteria is according to the sources of their energy. The photosynthetic bacteria derive energy from light and use carbon dioxide and mineral salts for growth. Photosynthetic bacteria so far have not found much use in bioconversions. Autotrophic bacteria obtain energy by means of the oxidation of inorganic substances, such as sulfur, hydrogen, or ammonia. They also derive their carbon from carbon dioxide. This is another class that has not been useful for bioconversions. The third and largest class is that of the heterotrophic organisms. These obtain their energy from the oxidation of organic compounds. Usually the carbon and sometimes the nitrogen and sulfur necessary for growth are obtained from the breakdown products. Most of the bacterial bioconversions are performed by the heterotrophic organisms. It is this group upon which we will concentrate our descriptive efforts.

Superimposing the Gram reaction distinction and the obvious differences in cellular morphology on the energy source classification, main divisions of the heterotrophs can be described as follows: Gram-positive cocci (spheres), Gram-positive rods (cylinders), Gram-negative rods, Gram-negative cocci, and Gram-negative curved rods and spirals. The Gram-negative rods are subdivided into an aerobic oxidative group, and an anaerobic fermentative group. It is implicit in this classification that the bacteria appear in three basic forms: cocci, rods, and spirals. Another structure is that of branched rods. These occur mostly in the Gram-positive group, and there is a regular progression of increased branching among these species until one finds completely filamentous forms. These are the actinomycetes which would appear to be fungi, were it not that they are prokaryotic and their size is much too small. The actinomycetes, especially the streptomycetes, are an important group of organisms used for bioconversions.

The organism's metabolic relationship with oxygen is another important factor in the classification of heterotrophs (as well as other microorganisms). Aerobic organisms live and thrive in an oxygen atmosphere. Oxygen is their ultimate electron acceptor. Anaerobic organisms thrive without oxygen. For electron acceptors they employ inorganic salts, such as nitrate and sulfate, or oxidized organic compounds, such as carbon dioxide. Certain anaerobes (''obligate anaerobes'') are killed by exposure to oxygen. Others that grow either in the pressence or absence of oxygen are the facultative anaerobes. These are the main divisions with regard to oxygen relationships. The term, ''microaerophilic'', occasionally turns up; this refers to organisms that seem to be benefitted by small amounts of oxygen, less than that present in the atmosphere.

In summary, the main types of organisms used for biotransformation of organic compounds are bacteria and the higher fungi (including the yeasts). The reader by this time should have the beginning of a long and profitable acquaintanceship with these useful microorganisms. It is interesting to speculate why it is only these kinds of organisms that have been used by and large for bioconversion studies. Perhaps the reason that others have not found so much use lies with the routine methods used for culturing of microorganisms. Could it be that only those organisms that are cultured conveniently with the present techniques are the ones that are used in the biotransformation schemes? It would seem that the neglected species must have interesting and useful compliments of enzymes of value for bioconversion studies. The neglect of these organisms, however, has not hampered the progress of the technology and science of bioconversion. If anything, there is such an abundance of organisms from which to choose, as well as the probability that many more easily cultured and useful species lie somewhere yet undiscovered, so that the real problem becomes one of choosing among the enormous possibilities.

III. BASIC MICROBIOLOGICAL TECHNIQUES

A. Culture Acquisition

There are three ways to procure cultures. They can be obtained from other investigators and colleagues working in the bioconversion areas. They can be isolated from natural habitats, and they can be obtained from various culture collections. Many such collections are located around the world. A comprehensive listing has been published.[5] For most purposes cultures can be procured conveniently from two collections in the U.S. These are the American Type Culture Collection, 12301 Parklawn Drive, Rockville, Md. (Abbr. ATCC), and the collection at the Northern Regional Research Laboratories, U.S. Department of Agriculture, Peoria, Ill. (Abbr. NRRL). Some of the more heavily used other major collections abroad are listed by Kieslich.[6] A discussion of enrichment procedures for isolation of cultures from nature will be given later in this chapter.

B. Microscopy

It is wise to develop some facility with the use of the microscope when working with microorganisms. Microscopy provides a quick way to check for culture purity, and to make observations generally at the cellular level. For observation of bacteria, it is recommended that an instrument with phase contrast optics with about 1000× magnification be available. With this optical system it is not difficult to study bacteria in the living state. One simply places a drop of a culture suspension, suitable diluted, onto a microscope slide, covers with a cover slip, focuses, and observes. Motility (other than gliding movements) is easily spotted. (However, the number of flagella and their mode of attachment are best seen by means of electron microscopy.) To distinguish motility from Brownian motion one looks for purposeful nonrandom movements. With most motile bacteria this is quite obvious. If the organism is highly aerobic, motile cells should be looked for at the edge of the cover slip where the oxygen level is higher. To measure the size of objects, an ocular micrometer is used. The ocular micrometer is calibrated at the particular magnification with a stage micrometer. Most bacteria encountered in biotransformation studies have about the same size range for width or diameter (from 0.1 to 1.0 μ). The average length of rods ranges from about 0.5 to 3 μ or more. Filaments 50 μ or more in length may be observed. Cells, both cocci and rods, are single, or associated in chains and various kinds of clusters, such as pairs or tetrads. Other types of structures that are seen are bent cells, spiral cells, branched cells, and of course filaments in various states of branching. Bergey's Manual[3] and Skerman's Guide[4] are good sources of descriptive information regarding bacterial cell morphology.

For a more detailed examination of cell morphology it may be preferable to make a stained preparation. A simple stain with a rather general applicability is the methylene blue stain. The procedure requires first that a small drop of a culture suspension be smeared and dried on a slide. The air-dried cells then are fixed to the slide by careful heating over a small flame. Heating must be just sufficient to fix the cells to the glass surface without rupturing or burning them. The fixed cells can be bathed in a 0.3% solution of methylene blue in water for 20 to 30 sec, and then washed off with distilled water. The slide is allowed to dry after gently blotting with tissue paper. At this point the specimen is ready for microscopic examination. Standard bright field optics with oil immersion objectives are recommended. Further details can be found in the *Manual of Clinical Microbiology* which has an excellent chapter on basic microscopy.[7]

Fungal cultures are best observed with a stereomicroscope with low to medium magnification. Slide preparations with a small block of nutrient agar covered with a coverslip provide standard chambers for growing fungi in a way that facilitates microscopic examination. Before covering, the agar block is inoculated on all four sides. The slide is incubated in a petri dish that contains a piece of wet filter paper on the bottom to insure that the slide does not dry out. With this mount the culture can be observed at various stages of growth without being disturbed. One of the important factors used for the identification of fungi is the observation of vegetative spore morphology, color, and arrangement. Another is the observation of the sexual spore arrangement and morphology. Useful reference guides for the identification of fungi are *The Fungi,* edited by Ainsworth et al.[8] and Stevens' *Mycology Guidebook.*[9]

C. Aseptic Techniques

Of all the techniques employed for microbial transformation of organic compounds, the mastery of the methods for the growth and manipulation of microorganisms is fundamental to success. The first step in a bioconversion study is the procurement of a suitable culture. Cultures usually are received in a vegetative state on nutrient slants or preserved by lyophilization in vials. Upon receipt these organisms must be subcultured and then preserved. In order to subculture, the investigator must learn aseptic technique, and how to obtain or make microbiological growth media. For work in bioconversions, and indeed in all areas of microbiological research, pure cultures are essential. These can be maintained and grown properly only through the application of techniques that do not compromise sterility. These are the first methods that must be mastered. In doing this, the first principle to be embraced is that no surface or substance may be considered sterile unless it has been sterilized and expressly kept that way. Microorganisms and spores are lurking everywhere. This includes the air in the laboratory. Even carefully washed hands are not sterile. Therefore, media and media vessels must not be touched after sterilization in any place where contamination is not wanted.

1. Sterilization

The best and most generally useful method for sterilization is by the use of the high pressure steam autoclave. Steam at 15 to 20 psi is required. For sterilization of relatively small amounts of material it usually is sufficient to maintain a temperature of 121° with 15 psi of wet steam for 15 to 20 min. Sterilization with dry heat requires more time and higher temperatures. Examples of "small amounts" would be several racks of test tubes containing 10 mℓ of medium, or ten 125-mℓ culture flasks with 25 to 50 mℓ medium in each. Amounts substantially larger should be sterilized longer, perhaps 30 to 40 min, but it should be pointed out that lengthy exposure to high temperatures destroys nutrient medium components. For sterilization of liquids the release of pressure must be controlled so as not to boil the liquids from the vessels. For nonliquid items it usually is best to dry them in the autoclave after sterilization. Both modes of sterilization are accomplished conveniently by the autoclaves currently available. It is highly recommended that a modern autoclave be available in the laboratory or nearby. While the equipment is rather expensive, to attempt sterilization in pressure cookers or by other means is awkward, and can result in failures as well as an overall waste of valuable time.

Liquids must not be autoclaved in sealed vessels. Several microbiological closure systems have been developed that function very well during autoclaving, and afterward during the culture incubation. A satisfactory closure must allow free exchange of metabolic gasses, while barring the entry of contaminating organisms. For glass culture tubes (sized 18 × 125 mm), disposable plastic closures are available. Glass disposable culture tubes also are available. If it can be afforded, disposable items are recommended for use whenever they can be obtained. Use of such items saves time and also can reduce cross contamination of cultures. For flasks the standard closure used for many years has been the surgical cotton plug. A variation on this is the cheesecloth cotton sandwich that is fitted over the top of the flask and held in place with a rubber band. Milk filter discs work admirably for this purpose. These discs, which are about 165 mm in diameter, fit on 125-mℓ and larges flasks, even the wide-mouthed 2.8-l Fernbach flasks. The milk filter discs come packed doubled, and this is the way that they should be used.

When screw-topped test tubes or vials are sterilized the caps should be placed on loosely, and then after sterilization and cooling to ambient temperature, the caps can be tightened. Screw-topped tubes are useful for long-term storage of cultures on agar media in the refrigerator where drying out of media with the more open plastic closure sleeves will occur. All autoclaving procedures should be done in trays because spills seem to be inevitable. Cleaning trays is easier than cleaning an autoclave. Utensils, such as spatulas, spreading rods, toothpicks, pipettes, syringes, and the like are autoclaved wrapped in aluminum foil. They should be wrapped so that they can be opened in a manner that minimizes contamination from airborne particles. Whenever possible, it is advisable to buy sterilized, disposable items such as pipettes and petri dishes. Disposable culture bottles of various shapes, and even disposable plastic spreading needles and loops are available. However, for the most part glass culture flasks are washed and reused.

2. Transfer

It is one thing to sterilize the materials needed for culturing; it is quite another to preserve sterility during the manipulations required for subculturing, transferring, and maintaining pure cultures. The transfer of cultures, termed inoculation, is a seeding process whereby small amounts of vegetative cells or spores are added to a sterile growth medium. This seeding must be done without introducing foreign microorganisms. A sterile hood with laminar air flow is very useful for making transfers, but it is not essential. Techniques that employ an open flame are quite adequate, and in some cases might be preferred, especially when culturing organisms that produce airborne spores in abundance. Such spores are difficult to control in a laminar air flow, and can be an annoying source of contamination.

The open flame technique can best be learned from an experienced microbiologist. Nevertheless, we shall give the reader a preliminary exposure to it. The principles to remember are that the flame is used to burn off contaminants from the transfer needle or loop, and to heat air at the neck of the flask or tube so that air flow is outward, not inward. Also, the flame is used to provide a hot, sterile curtain above the open culture vessel. No openings are made without the presence of the flame. When tops are removed, as they must be before flaming, they are held so that their sterile surface is pointed downward. Openings are for the shortest time possible to minimize the chance entrance of contamination. When transfers are made from the living culture mass on the surface of a slant or plate, a metal transfer loop is used. These are readily available from equipment supply houses. Between each cycle of touching to the culture surface and transfer to the new medium, the loop must be heated to redness, and then cooled by touching it to a clean area of agar surface without growth, or by dipping it into sterile nutrient medium.

When transferring from slant to liquid media, a small amount of culture again is scraped up with the loop, and then agitated in the fresh medium until the seed is washed off the loop. Even when transferring the same culture to a series of flasks, the needle must be burned off, and cooled with each operation. An open flame does not work so well in a laminar air flow hood; in this case it is preferable to sterilize needles and loops with a small electric oven built expressly for this. When transferring from one liquid culture to another, the microbiological loop also may be used. Otherwise, several drops of the inoculum culture fluid may be transferred conveniently with a sterile pipette. When transferring from tube to tube it is possible to hold both tubes between the fingers of one hand, remove closures with the other hand, holding them upside down, while making the required transfers and flamings. For larger scale transfers one simply pours from one vessel to the other, while taking care to flame openings, if not working in a sterile hood. The sterile hood makes many operations easier and more reliable (except where spores are a problem), but it still is a place where great care must be exercised to prevent contaminations.

Inoculation of solid media requires practice, but basically the operations are simple. The three most common solid media are slants (slopes), stabs, and plates. Making a stab into a nutrient agar plug in a culture tube is the simplest task. The needle (suitably flamed and cooled) is dipped into a liquid culture, or scraped lightly over the surface growth on a slant or plate, or even just touched lightly to the tip of a colony on a plate, and then stabbed into the agar in the tube. Usually two stabs are made side by side before flaming the needle for the next operation The needle should be kept steady on the way in and out. Depending upon the oxygen requirements of the organism, growth will occur down the length of the stab, deep down, or only near the surface. Stabs are useful for maintenance of facultative anaerobes.

The inoculation of an agar slope is accomplished after picking up organisms on the needle as described above, by drawing the needle carefully up the length of the slant, while making an oscillatory motion. If it is desired to have a more confluent growth on the slope, the inoculation should be done with a loop, and as the loop is drawn upward the surface of the slant is "painted" with the loop, taking care not ot break the gel surface. The same action is required when a loop is used to inoculate the surface of an agar plate. This operation is referred to as "plating". Plating has several purposes; one simply is to obtain a large area of confluent growth. Another is to isolate colonies, either to check for purity, or to purify the culture. This last is the single most important use of plating, for this is a reliable method to propagate strains from single cells. Plating also can be used to provide counts of viable cells in a given culture. It is important to become proficient with the plating technique.

When working outside a hood, the cover of the petri dish is never completely removed; rather it is raised just far enough to allow the insertion of the loop. Once inserted, the loop is spread over the surface of the agar gel with a sweeping or painting motion, so that the surface of the loop is gradually wiped clean of cells. In this way cells can be spread thinly enough so that regions occur where individual colonies can grow up from individuals deposited on the surface. There are numerous methods of plating for single colonies. One sure way is to start with a diluted culture so that the degree of separation is not crucial. Another way is to start the spreading process at the top of the plate, making this area well covered with cells, and then without changing the angle of the loop to the agar surface, painting wide, sweeping strokes over the rest of the surface, while being careful never to retrace strokes. Of course, as nature would have it, not all microorganisms will produce small, discrete colonies. Some bacteria, and especially fungi, have a spreading mode of growth and tend to cover the entire surface. By the addition of inhibitors, such as oxgall, to the media, it sometimes is possible to obtain discrete colonies from spreaders.[10]

For making viable cell counts, a reliable method is to plate 0.1 mℓ and 1.0 mℓ of each of the samples of serial tenfold dilutions of a culture. The spreading is this case is done suitably with miniature glass hockey sticks. Spreading is facilitated when the plates are rotated automatically. The pour-plate method is another way to determine viable counts. According to this technique, 0.1 mℓ of each of the dilutions is added to 10 mℓ of agar medium in tubes held at 45° (just above gelling temperature), mixed well, and then poured into sterile petri dishes. During incubation, colony formation takes place mainly within the gel.

The transferring of fungal cultures requires special discussion. For vegetative transfers of surface growth it is possible to scrape up fragments of mycelia with a loop. Plugs of growth can be obtained with loops or sterile straws. If the organism sporulates profusely (or can be encouraged to by special growth media), so much the better, for now spore suspensions can be made, and these are very convenient. Usually spores can be removed from surface growth by washing with dilute surfactant solution (0.01 to 0.1%). A suitable surfactant is Tween 80, one that is relatively nontoxic for most species. The surfactant suspension can be used directly as inoculum, or the spores can be collected and washed by centrifugation, and then resuspended in distilled water. Such suspensions can be calibrated. When stored in the refrigerator, the viable spore counts remain relatively constant for months at a time. Especially in small flasks, fungi tend to grow so that the mycelia conglomerate. This is disastrous when the cultures are to be used for inocula. Short treatments (30 to 60 sec) with a blender or sonifier (both sterilized) help to break up these clumps, allowing the subsequent outgrowth of the culture to be more uniform.

D. Growth Media for Routine Use

For the routine culturing and maintenance of the culture collection it is convenient to use the prepared media that can be purchased. The procedures for constructing specialized microbiological growth media will be given in subsequent paragraphs. The names of suppliers of packaged media are Difco Laboratories, Inc., Detroit, Mich., and Oxoid Ltd., Southwark Bridge, London, SEI. For the most part these media are premixed and dehydrated. To prepare them it is only necessary to add water, stir to dissolve, dispense into containers, and sterilize. Certain media also have agar already added, but if agar is needed it should be added at 1.5 to 2.0% (w/v). Usually agar must be dissolved by heating carefully to boiling while stirring, although some types will dissolve and disperse nicely during autoclaving. One such agar is oxoid Agar No. 3.

Most bacteria can be grown in Nutrient Broth (Difco) which consists of beef extract and peptone. Another all-purpose medium is Dextrose Broth (Oxoid); this contains glucose along with meat extract and a tryptose. M.R.S. Broth (Oxoid) is useful for the cultivation of faculative anaerobes, especially lactobacilli. This is a medium based on glucose, yeast extract, meat extract, and peptone. Recommended for the cultivation of anaerobes are Reinforced Clostridial Medium (Oxoid) which contains cysteine for oxygen scavanging, and Thioglycollate Medium (Difco) which contains both thioglycollate and cysteine. A medium consisting of glucose (1%) and yeast extract (Difco, 1%) serves well for the growing of a wide variety of bacteria (including actimonycetes), yeasts, and molds.

For the growth of fungi, corn meal agar, potato dextrose, Czapek Dox, and Sabouraud Dextrose media have been used extensively. These are available from either company listed. Czapek Dox medium is the only one that contains sucrose for the carbon source. All the others contain glucose, except the corn meal medium which has nothing added. YM medium (Difco) is especially suited to the cultivation of yeasts. This is a complex medium with glucose once again as the carbon source.

A variety of microbiological media can be purchased already sterilized and dispensed in plates, slants, and tubes, but it is usually just as convenient to make up one's own media in advance, and store it until ready for use. Routine cultivation procedures require that agar-gelled media should be available as plates and slants, and that liquid (broth) media be available in tubes and flasks. Slants are prepared in 18 × 150 mm test tubes containing 10 mℓ of suspended agar medium; the tubes capped with closures are autoclaved upright in racks for 15 to 20 min. After about 10 min cooling in air, each tube is mixed with a vortex mixer, and then propped against a rod for cooling and solidifying in the slanted position. Forty or more slants of the various media can be prepared at a time, but the number prepared should be such that most will be used before they dry out. Slants should be good for 2 to 3 weeks in the ordinary laboratory environment.

Agar media for plates must be autoclaved separately and then poured into sterile petri dishes. The common, sterile plastic dishes are 100 mm in diameter, and hold approximately 20 to 30 mℓ of medium. Pouring plates in a sterile hood is no problem. Without the benefit of a hood, a good technique is to open them up as a clam shell while pouring the agar, and then immediately replace the lid. If the agar medium is cooled to about 50 to 60° before pouring, the operation is easier and less condensation forms on the petri dish lid. Twenty to forty plates should be prepared at a time for convenience. Sealed up in their original plastic bags to retard evaporation, these plates keep in a refrigerator for several months.

Broth media in test tubes are useful for starting new cultures and propagating facultative anaerobes; usually 10 mℓ per tube. While sterilizing tubes of media, it is prudent to prepare test tubes of distilled water and 0.9% saline solution in 9 mℓ amounts. These are useful for various reasons, such as making dilutions for plating or microscopic investigation. Erlenmeyer flasks (250 mℓ) containing 25 to 50 mℓ of medium are excellent for cultivation of bacteria and yeasts. These have proved to be a superior combination of size and volume of medium both for convenience and aeration efficiency. For the filamentous fungi, larger flasks such as the 500 mℓ erlenmeyer with 100 mℓ of medium allows for a more uniform dispersion of growth and provides a larger volume from which to sample. The flasks are closed with milk discs or cotton plugs as mentioned above. Covered with aluminum foil to retard evaporation, these flasks can be stored several weeks in the refrigerator.

E. Subculturing New Cultures

When a new culture arrives it must be grown up in an appropriate broth, examined microscopically, and plated out to check for purity. If the sample is lyophilized, break open the vial aseptically. Then add 1 or 2 mℓ of broth medium to resuspend the culture. This suspension is used to inoculate the growth medium; a loop also should be plated out. If the new arrival is on a slant, scrape off a small amount of vegetative growth and resuspend it in a broth. Also plate it out. Cultivate aerobes in 250-mℓ flasks containing 25 mℓ medium; shake at about 200 rpm at the proper temperature. Facultative anaerobes should be grown in tubes without agitation. A convenient way to cultivate strict anaerobes is to use small (50 mℓ) glass-stoppered bottles filled completely with medium. After inoculation, medium is poured into the neck of the bottle, and then the stopper is dropped in to make the seal. These bottles are best incubated in trays to catch liquid that might be expelled during incubation by the evolution of gas. Anaerobes also can be grown on plates in special anaerobic jars available from several manufacturers.

F. Culture Preservation

After a new culture has grown up it must be plated out and examined for purity; it may be necessary to repeat the plating process to be assured of having pure colonies. A single colony of the culture is picked, and transferred to slants or stabs. Slants are the best way to store most aerobic stocks, including yeasts and fungi, for periods of several weeks. Facultative anaerobes store best in deep stabs. Anaerobes can be stored sealed in tubes. After inoculation, each of these various agar media should be incubated for several days or until growth is abundant, and then they may be stored in the refrigerator for several weeks or months depending on the species. When reactivating these organisms, it is good practice to examine them microscopically, and to plate them out for individual colonies.

For long-term storage, the best two methods in terms of general applicability are liquid nitrogen freezing, and lyophilization.[11] If a supply of liquid nitrogen is available, the most convenient preservation method is the ultra-low temperature storage. Many species survive liquid nitrogen storage better than lyophilation. The liquid nitrogen method requires that a culture be grown up maximally, collected by centrifugation, and then resuspended in sterile glycerol solution (10%). Some workers prefer dimethylsulfoxide solution (10%). The suspension is immersed in liquid nitrogen for freezing and kept that way until ready for reactivation. Sterile plastic vials, as well as storage systems, are available for small to large collections. Thawing should be done rapidly in a water bath at 40°.[12]

The lyophilization procedure also involves growing up the culture maximally, harvesting by centrifugation, and then resuspending the cells in sterile skim milk. Care must be taken to ensure that the milk has been completely sterilized, for it frequently contains bacillus spores. The cell suspension in milk is dispensed into small sterile vials that fit onto a high vacuum system. After freezing in a dry ice-acetone bath, the cell and milk mixture is sublimed at room temperature under high vacuum. When the drying is finished the vials are sealed under vacuum. Such preparations are viable for long periods.

A check for viability is required for each of these procedures. Although the survival rate from the liquid nitrogen freezing is higher, it is of course possible that either method will be unsatisfactory for a particular species. If so, other preservation methods must be sought. A good discussion of preservations can be found in Perlman's chapter in *Application of Biochemical Systems in Organic Chemistry.*[11]

IV. MICROBIAL GROWTH AND NUTRITION

A. Growth Kinetics

Growth of microorganisms is the development of cell mass at the expense of nutrients; it is the proliferation of cells used for the inoculum. Growth of filamentous organisms results in the increase of mycelial mass. On solid media, such as plates, growth appears as a mass of substance on the surface. On plates, properly seeded growth can appear as single colonies distinctive in shape. The single colonies arise in areas where single cells were deposited during the plating operation. Some fungi and bacteria form large colonies that can spread over most of the surface, but many microorganisms form colonies small enough so that 50 to 200 can be accommodated easily on a 100 mm plate. The size, shape, surface texture, and color of colonies are important characteristics of the different species.

Growth of bacteria and yeast in liquid media appears as turbidity which increases during the growth cycle until the medium becomes completely opaque. In the case of bacteria, the turbidity will appear as a streaming fluid when the culture is swirled and held up to the light. Mycelial growth is less homogeneous, frequently occurring in balls, clumps, and stringy masses. In broth, microorganisms undergo a cycle of growth phases. Starting from a small inoculum, the phases are lag, exponential (logrithmic), stationary, death, and survival. These are diagrammed in Figure 1. These kinetics result from the process of reproduction by binary fission, or the budding type of reproduction.

The lag phase occurs because it is necessary for the cells to adapt to their new physical and nutritional environment before starting to grow rapidly. The exponential phase occurs when growth by binary fission is at the maximum rate for the prevailing conditions in the medium. The stationary phase starts when nutrients begin to run out, or when metabolic waste materials begin to inhibit. After a time in this phase, cells begin to die off, signaling the start of the death phase. At the end of this phase, there

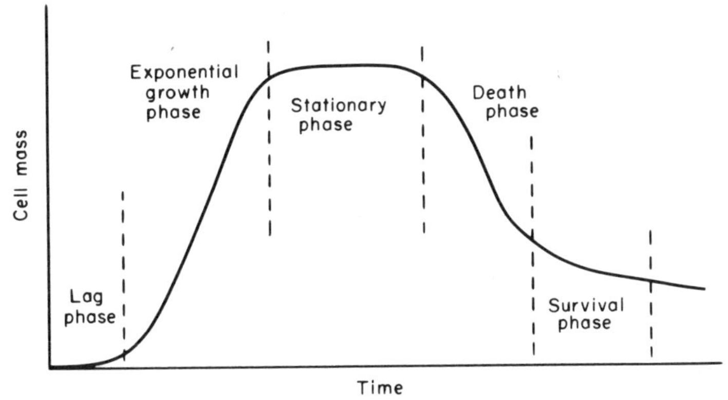

FIGURE 1. Bacterial growth phases.

may be a survival phase in which some proportion of the cells manage to survive for relatively long periods. Growth kinetics are affected by physical environment and nutritional environment. These effects will be discussed below. It should be pointed out that organisms that reproduce and grow by means other than budding and binary fission will exhibit different growth kinetics. The most noticeable difference is in the log phase.

B. Measurement of Growth

There are several different ways to measure growth. In bioconversion studies it is necessary to know the cell mass expressed as dry weight. Dry weight is determined by drying a homogeneous sample of a culture throughly and weighing it. Such a measurement of dry weight is rather tedious, and so a short cut that works well for homogeneous cultures is to calibrate cell dry weight against turbidity under standard conditions. This is done by preparing a thick suspension of cells, and drying an aliquot (duplicate, at least) for weighing. Following this, the suspension is diluted appropriately to give about six to eight different densities in the region between 0.1 and 1.0 absorbance (usually at 650 nm to reduce effects of yellow colorations). The absorbance is plotted against the dry weight calculated proportionally for the particular dilutions. A straight line falling off somewhat in the more turbid region is obtained.

If viable cell numbers are required, a plate count is necessary. This is done by the serial dilution method. Dilutions of $1/10^2$ are made serially by placing 1 mℓ of a cell suspension into a flask containing 99 mℓ water or 0.9% saline. After thorough mixing, 1 mℓ of this dilution is added to another flask containing 99 mℓ. This operation is repeated three more times to obtain five flasks that contain in successive order dilutions of $1/10^2$, $1/10^4$, $1/10^6$, $1/10^8$, and $1/10^{10}$. Each of these is plated twice on agar, once with 1.0 mℓ, and once with 0.1 mℓ, resulting in a series of 10 plates with dilution factors ranging from 10^2 to 10^{11} per mℓ. Counts of cells also can be made microscopically by means of a calibrated chamber (e.g., a hemocytometer), and also by proportional methods. For instance, it is known that blood contains on the average 10^9 red cells per mℓ. When equal amounts of blood and a bacterial suspension are mixed and spread on a plate, it is found that the number of bacteria divided by the number of red cells times 10^9 (times the dilution factor) gives the number of bacteria per mℓ. Of course, only the plate method will give numbers of viable cells. The viable cell count usually runs below the total count. As a handy rule of thumb, a suspension of an average size bacterium, such as *Escherichia coli,* that is just visible as turbidity to the naked eye contains about 10^6 cells per mℓ.

The measurement of mycelial cell mass is more tedious. Of course, drying and weighing samples is the direct way to do it, but frequently there is difficulty in taking homogeneous samples. By the adjustment of medium and growth conditions, it may be possible to promote a more dispersed growth. Failing this, one may blend the entire contents of a flask before sampling. Other methods of quantifying growth are discussed in a chapter by Calam in *Methods in Microbiology.*[13]

C. Effects of Environment

1. Temperature

The temperature range in which life can exist almost approximates the range in which water is in the liquid state. Growing life forms are observed at temperatures from −5° to about 80°, but no one organism can thrive over the entire range. Rather, organisms have adapted for growth in three distinct temprature ranges. The mesophiles live at temperatures from 10° to 47°; they thrive at 30° to 45°. This is the most abundant group, and the majority in this group prefer a range from 27° to 37°. Psychrophiles prefer lower temperatures, living and thriving as low as −5° and up to about 35°. Their temperature optimum, if they are obligate psychrophiles, is 15° to 18°. Facultative psychrophiles prefer somewhat higher temperatures (25° to 30°). On the other end of the scale are the thermophiles, which will grow over a range of 40° to 80°, their optimum being from about 55° to 75°. Thermophiles are interesting because, among other things, they possess enzymes with remarkable heat stability.

Throughout each of the three temperature ranges, the organism's growth rate varies with the temperature. Except at the extremes, the rate of growth generally follows Arrhenius' Law. This owes to the fact that growth is dependent upon enzyme-catalyzed chemical reactions; growth can be viewed as a chemical process. Furthermore, specific bioconversion reactions can have their own optimum temperature, which is not necessarily the optimum temperature for growth. Bioconversions conducted with non-growing cells or spores can take advantage of this fact.

2. Oxygen Requirements

The organisms that require molecular oxygen to live and grow are called aerobes. Certain species cannot grow in the presence of air and indeed are killed by oxygen; these are the obligate anaerobes. Still others are able to thrive in the presence or absence of air; these are the faculative anaerobes. Some aerobes grow best when the oxygen partial pressure is less than normal. These are termed microaerophiles. Highly aerobic organisms usually will grow best when their medium is aerated maximally. Such conditions may also be best for the bioconversion process, if it is linked to growth. If not, the bioconversion process could show its own aeration optimum. Perhaps the most crucial factor in the cultivation of aerobic organisms is the aeration efficiency that can be achieved. This efficiency is measured as the oxygen transfer rate, or the mass transfer coefficient. In flasks an acceptable oxygen transfer rate can be achieved in 250-mℓ flasks, containing 25 mℓ medium, shaken at 200 rpm on a rotary shaker with a 2 in. orbital throw. This efficiency is almost matched in 2.8-1 Fernback flasks with baffles, containing 500 mℓ medium, shaken at 150 rpm.

Although facultative anaerobes do not require air, they usually achieve better growth rates when they are aerated. The growth of strict anaerobes must be conducted in the absence of oxygen. In nature these creatures are found in environments rich in sulfide. Thus, the addition of sulfides or thiols to their growth media approximates natural conditions. Sodium sulfide at a concentration of about 0.05% is sufficient. Other effective reducing agents are cysteine and sodium theiglycolate at about the same concentration as sodium sulfide. Most media can be autoclaved with the thiols included, as long as they are not acidic. As mentioned previously, anaerobes are cultured easily in completely filled, glass-stoppered bottles. Alternatively, the anerobic jar can be used.

3. Acidity

Most microorganisms grow best at pH values between 6.5 and 7.5. For this reason phosphates usually are the buffers of choice in media. Phosphates are innocuous, and although they do supply a required nutrient, they are more important as buffers. Outside of the phosphate buffering range, bicarbonate can be used for higher pH values, and acetate or dimethylglutarate can be used at the lower values. Dimethylglutarate generally is not metabolizable. Bacterial species, for the most part, do not grow at pH values much below 4.5, or above 9.0. This includes the streptomycetes. As would be expected, there are exceptions. *Thiobacillus thiooxidans* grows at values in the range of pH1. Certain *Bacillus subtilis* strains grow above pH 9.0. These are a few of the interesting examples. Although organisms will stop growing at given acidities, they well might survive these conditions. Furthermore, at unusual pH levels certain useful enzyme activities may be enhanced, providing yet another factor that could be employed to advantage in bioconversions with nongrowing cells. Molds and yeasts grow well at pH values 3.5 and up to 7 or 8, but generally they prefer the range of 5 to 6. Again, it must be recognized that we are dealing with generalizations, and that exceptions will occur.

4. Osmotic Pressure

Ordinary microbes are inhibited by 2 to 5% sodium chloride. Salt-tolerant, and indeed, salt-loving organisms exist; these are the halophiles. Moderate halophiles grow in 2 to 20% salt. Extreme halophiles require at least 10% salt and tolerate up to 31%. These salt-loving organisms have developed ways to prevent water from flowing out of the cell into the medium under conditions of extreme osmotic pressure differentials. Furthermore, certain yeasts are able to grow in 70 to 80% sugar solutions. Interestingly, these yeasts are not especially salt tolerant, and conversely, the halophiles are not adapted to grow in the high sugar concentrations. The osmotic pressure of the nutrient solution for most microorganisms should be less than 5 atmospheres for the Gram-negatives, and less than 20 atmospheres for the Gram-positives. These values must be kept in mind when constructing growth media.

Water is required for life and growth, of course. The lower limits of available water that support growth of yeasts and bacteria are of the order of 18 to 20%; fungi can do with a little less, about 15%. These values are applicable only to growth, for many organisms remain viable in a semidry state. This is the reason why drying methods and particularily lyophilization work so well for the preservation of cultures. Lyophilization is a very gentle way of removing most of the free water within the microorganism.

D. Media Construction

Water constitutes 80 to 90% of the weight of cells. The remaining solid matter is made up largely of the elements, carbon, oxygen, nitrogen, hydrogen, phosphorus, and sulfur in the respective percentages of 50, 20, 14, 8, 3, and 1. Approximately 96% of the dry weight is accounted for by these elements. The remaining constituents are made up of many elements, among which are sodium and potassium (each at 1%), magnesium and chlorine (each at 0.5%). Next in abundance is iron (at 0.2%), and then follows a group of trace elements, including manganese, cobalt, zinc, molybdenum, and copper, all of which in total represent only 0.3% of the dry weight. These data were assembled by Luria for *Escherichia coli.*[14]

In the construction of a nutrient medium for microorganisms, ingredients must be added that can serve as sources for the various required elements. Traditionally, microbiological media were made up of extracts of plant and animal tissues, for obviously, these materials should contain many of the nutrients necessary for the maintenance of life and growth. In addition to sources for the required elements, the organism requires an energy source. The organisms with which we are concerned mainly in bioconversion research are the chemoheterotrophs, which obtain their energy by the oxidation of reduced organic compounds. For aerobic organisms the ultimate electron acceptor in oxidation is oxygen. The anaerobes utilize a variety of electron acceptors, including the inorganic ions, nitrate and sulfate. Various organic compounds, such as carbon dioxide, likewise accept electrons.

It would be safe to state that there is no naturally occurring organic compound that cannot be used as a carbon source by some microorganism. The selectivity for energy sources is more stringent. In addition, there are many of the unnatural, synthetic organics that can be used as carbon sources. This phenomenon is of particular interest in the biotransformation of organic compounds by microorganisms, for it means that enzymes exist that can perform reactions on a wide variety of substances. Organic compounds also supply oxygen and hydrogen to cells (water is another source). Nitrogen and sulfur can be supplied either in organic or inorganic forms. Many organisms can assimilate ammonium, nitrate, and sulfate salts, with special capabilities varying with the species. Some require additional organic growth factors, including amino acids, peptides, vitamins, purines, and pyrimidines. Some growth factor requirements are even more complex than that. It will be appreciated that media containing extracts of yeast, plants, or animals will contain many of these required substances, including the more complex ones.

The aim in the construction of a medium is to supply all the required nutrients in amounts that warrant good growth, but do not inhibit. Concentration is very important; usually there is an optimum range that promotes very good growth. Beyond this range there is inhibition. Another source of inhibition is the accumulation of toxic metabolites during growth. These metabolites may just affect the pH value of the broth so that it is out of the optimum range. Growth on glucose, other carbohydrates, and alcohols usually results in the accumulation of organic acids. The consequent lowering of the pH value can be controlled somewhat by increasing the potassium phosphate buffer concentration to 0.5%. Beyond this concentration phosphate can be inhibitory. The addition of calcium carbonate at 0.5% or more is an effective way to keep the pH value near neutrality when acids accumulate. The resulting pH value is usually between 6.7 and 6.8. Unfortunately no equivalent, nontoxic, insoluble acid exists for controlling the rise of pH values in medium. Another way to control acidity is by means of periodic aseptic additions of acid or base during the growth cycle. This can be awkward.

Crowding also can slow the growth rate through competition for nutrients and other factors. For shaken flasks, the crowding of itself rarely is severe enough to significantly slow growth. What usually happens in the culture of aerobes is that the dissolved oxygen concentration in medium falls to a limiting point.

Two types of media commonly employed are the synthetic medium and the complex medium. When the constituents are chemically defined, the medium is called synthetic; if extracts or digests of plant or animal tissues are used, the medium is called complex. The construction of a synthetic medium starts with the addition of a basal mineral salts mixture which is the source of the trace elements and other salts. There are many examples of salt solutions in the literature. One that is suitable for routine use is given in Table 1.[15] In synthetic media, nitrogen is supplied by ammonium sulfate, ammonium nitrate, or potassium nitrate. Some species cannot assimilate nitrate. Other defined nitrogen sources are urea and the amino acids, aspartate, asparagine, and glutamate. Sulfate supplies the sulfur, in most cases. The concentration of the inorganic nitrogen sources should be about 0.1 to 0.2%; organic sources can be used at about the same concentrations. Buffering capacity as well as a donation of phosphorus is provided by potassium phosphate. Its concentration should be in the range, 0.1 to 0.5%.

Table 1
COMPOSITION OF A SALT SOLUTION FOR USE IN DEFINED MEDIA

Magnesium sulfate · $7H_2O$	25.00 g
Ferrous sulfate · $7H_2O$	2.80 g
Manganous · H_2O	1.70 g
Sodium chloride	0.60 g
Calcium chloride · $2H_2O$	0.10 g
Sodium molybdate · $2H_2O$	0.10 g
Zinc sulfate · $7H_2O$	0.06 g
Hydrochloric acid, 0.1 N	1.00 ℓ

Note: The solution is 100-fold concentrated.

Table 2
COMPOSITION OF A TYPICAL DEFINED MEDIUM[a]

Isobutyric acid[b]	2 g
Ammonium sulfate	1 g
Dipotassium phosphate	1 g
Salt solution[c]	10 mℓ
Distilled water	990 mℓ

[a] Adjust to pH 7.0 with potassium hydroxide.
[b] Isobutyric acid is only an example: a variety of carbon compounds can be substituted for it.
[c] See Table 1 for salt solution composition.

For chemoheterotrophs the organic carbon donor also doubles as the energy source when it is oxidized. Concentrations of un-ionized carbon compounds for routine use are from 0.1 to 2.0%. Higher concentrations can be tolerated. At higher concentrations, ionized compounds such as organic acids are more likely to be inhibitory. Finally, in the making of defined media it is sometimes useful to add a small amount of yeast extract (0.02 to 0.05%) to enhance growth rate and reduce lag time. These small amounts of yeast extract do not disqualify the medium from being a defined one. The composition of a typical defined medium is given in Table 2. It is important to sterilize glucose and other carbohydrates separately in this medium, because autoclaving in the presence of phosphate forms growth inhibitors. One way to do this is to make up the glucose and salts components separately in double concentration, sterilize, and then mix one to one when cooled. Furthermore, to prevent the formation of insoluble salts during autoclaving, chelating agents such as ethylenediamine tetraacetic acid (at about 0.01%) can be added. Citrate is another good chelator for this purpose. For all media, the pH value should be checked before sterilization and adjusted if necessary. It also should be checked again after sterilization.

The exact components of a complex medium are not known in detail at the molecular level. A typical complex medium is shown in Table 3. Experience has shown that this glucose-yeast extract medium can be autoclaved up to 30 min without apparent adverse consequences. In regard to carbon and energy sources for media of all types, glucose is certainly one of the most commonly used, since many types of bacteria, yeasts, and molds can utilize it. Other carbohydrates, citric acid cycle intermediates, and organic acids also are quite useful. Growth on glucose represses the synthesis of certain en-

Table 3
COMPOSITION OF A TYPICAL COMPLEX MEDIUM

Glucose	10 g
Yeast extract	10 g
Dipotassium phosphate[a]	1 g
Salt solution[a,b]	10 mℓ
Distilled water	990 mℓ

[a] These additions are not necessary, although they do stimulate the growth of some organisms.
[b] Salt solution from Table 1.

Table 4
EXAMPLES OF COMPLEX NITROGEN SOURCES SUITABLE FOR MICROBIOLOGICAL MEDIA[a]

Nitrogen source	Derivation
Peptones	A group of water-soluble protein hydrolysates formed by acid or enzymatic (usually pepsin) hydrolysis. A variety of animal and plant proteins are used. Casein is a common source. Products differ in source of protein and the degree of hydrolysis.
Tryptones	Water-soluble hydrolysates made by the action of trypsin on various proteins.
Beef extracts	Water extracts of ground beef. The methods of extraction allow varying degrees of bacterial action which result in some protein hydrolysis.
Yeast extracts	Water extracts of autolysed yeast which contain not only water-soluble components, but also products of enzymatic action on yeast proteins and other constituents.

[a] Can be obtained in dehydrated form from media supply houses, such as Difco, Oxoid, and BBL (Baltimore Biological Laboratories).

zymes, but this repression can be avoided by switching to another carbon source. In complex media, the important sources of nitrogen and other elements are enzymatic digests of casein or other proteins, such as soy protein. These digests are known as peptones or tryptones. Other complex nitrogen sources are corn steep liquor, yeast extract, beef extract, potato extract, and so on. Table 4 indicates commercial sources of some of these materials. These digests and extracts not only supply the needed nitrogen, but they supply additional carbon, as well as sulfur, phosphorus, and the various other elements and growth factors that may be required.

When solid media are required, simply add microbiological agar at 1.5 to 2.0%. Most agars require prior dissolution before autoclaving. Agar should not be autoclaved in media that are excessively acidic, because hydrolysis will occur. At pH values of 5 or below, autoclave agar separately in double concentration. For very acidic media use silica gel for the solidifying agent. Examples of complex media used for the cultivation of microorganisms of the various types common to a culture collection are given in Table 5.

E. Enrichment Culture

Through the enrichment culture method it is possible to isolate many different types

Table 5
SELECTED STANDARD MEDIA FOR CULTURING A VARIETY OF MICROORGANISMS[a]

1. Nutrient broth

Beef extract	3.0 g
Peptone	5.0 g
Distilled water	1.0 ℓ

For solid medium add 15 g agar.
pH 6.8 after autoclaving

2. Reinforced clostridial medium

Yeast extract	3.0 g
Beef extract	10.0 g
Peptone	10.0 g
Solbule starch	1.0 g
Glucose	5.0 g
Cysteine hydrochloride	0.5 g
Sodium chloride	5.0 g
Sodium acetate	3.0 g
Agar	0.5 g

pH 6.8
For solid medium increase agar to 15 g.

3. M.R.S. lactobacillus broth

Peptone	10.00 g
Beef extract	8.00 g
Yeast extract	4.00 g
Tween 80	1.00 mℓ
Glucose	20.00 g
Dipotassium phosphate	2.00 g
Sodium acetate · $3H_2O$	5.00 g
Triammonium citrate	2.00 g
Magnesium sulfate · $7H_2O$	0.20 g
Manganese sulfate · $4H_2O$	0.05 g
Distilled water	1.0 ℓ

4. Mannitol Broth

Yeast extract	5.0 g
Peptone	3.0 g
Mannitol	25.0 g
Distilled water	1.0 ℓ

Solidify with 15 g agar, if needed.

5. Yeast extract tryptone medium

Yeast extract	2.5 g
Tryptone	5.0 g
Glucose	1.0 g
Distilled water	1.0 ℓ

6. Potato dextrose agar

Potatoes, infusion from	200 g
Glucose	20 g
Agar	15 g
Distilled water	up to 1ℓ

pH 4.5
To inhibit bacterial growth, cool medium to 50° after autoclaving, and add 1.0 mℓ sterile 10% tartaric acid per 100 mℓ of medium. Final pH will be 3.5.

Table 5 (continued)
SELECTED STANDARD MEDIA FOR CULTURING A VARIETY OF MICROORGANISMS[a]

7. Malt extract broth	
Malt extract	17 g
Peptone	3 g
Distilled water	1 ℓ
pH 5.4	
8. Neurospora culture broth	
Yeast extract	5 g
Peptone	5 g
Maltose	40 g
Distilled water	1 ℓ
pH 6.7 after autoclaving.	
Add 15 g agar to solidify.	
9. Sabouraud dextrose broth	
Peptone	10 g
Glucose	20 g
Distilled water	1 ℓ
pH 5.7	
Add 15 g agar to solidify.	

[a] Media similar to these (and the ingredients that go into them) can be obtained in dehydrated form from Baltimore Biological Laboratories, Difco Laboratories, and Oxoid Limited, among others.

of microorganisms from their natural habitats. A helpful discussion of methods applied especially to the isolation of bacteria is found in *The Microbial World.*[16] The underlying prinicple of the methods is not complicated. Thus, in a given nutrient medium, the strains that proliferate the fastest will eventually predominate. This predominance can be exerted near the beginning of the culturing, or it can be attained gradually as the conditions in the medium change because of metabolic activity. For example, the shift to acidic conditions caused by one group of organisms ultimately will favor the strains that thrive in or at least tolerate, acidic media.

The variation of the nutritional components of a medium allows for the enrichment of different classes of microorganisms. If the only nitrogen source is nitrate, then nitrate assimilators will be favored. If nitrogen is left out entirely, nitrogen fixers will take over. If alanine is the only nitrogen source, strains that can metabolize alanine will predominate in the culture. By controlling the oxygen supply in the enrichments, one can isolate aerobes, facultative anaerobes, or strict anaerobes. It is obvious that when oxygen is plentiful, aerobic organisms will be abundant. When oxygen is limited, facultatives grow best; when oxygen is eliminated altogether, obligate anaerobes will dominate. In a rich medium with an easily fermentable carbon source, such as glucose, the oxygen level can be reduced to zero quickly by the action of the aerobes in the mixed population. From this point it is the turn of the facultative anaerobes and the strict anaerobes to multiply their forces. Without thiol compounds in the medium, facultatives will win the race. With thiols one can expect to obtain chiefly strict anaerobes.

To isolate yeasts and other fungi, the pH value of the medium should be on the acid side, perferably pH 5 or 6. The addition of penicillin G and streptomycin or dihydrostreptomycin (20 units of penicillin and 40 μg of either streptomycin per mℓ) to the

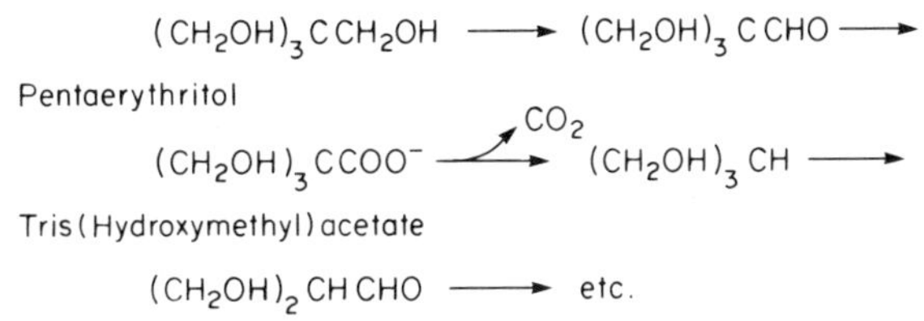

FIGURE 2. Microbial oxidation of pentaerythritol.

medium inhibits the growth of bacteria. Crystal violet (0.001%) also is an inhibitor of bacterial growth. The antibiotic must be sterilized by filtration and added to autoclaved medium that has cooled. If the medium contains agar, it is sufficient to cool to 45 to 50°. Littman Oxgall agar (Difco) is a choice medium for plating out fungi, especially if antibiotics are added. The medium already contains crystal violet. The oxgall inhibits spreading growth of the fungi, so that individual colonies form which are easy to isolate in pure culture.

The enrichment method of obtaining cultures has great value in biotransformation studies, for it is possible to find organisms that perform reactions on a great variety of organic compounds. A compound that is utilized as a carbon source is broken down to a series of metabolic intermediates by means of stepwise reactions. The creation of a metabolic block through mutation or manipulation of environmental or nutritional factors (or by the use of inhibitors) brings about the accumulation of the metabolite preceding this block. The bacterial oxidation of pentaerythritol provides an example.[17] The enzyme system of a *Corynebacterium* species degrades this compound in a stepwise fashion as shown in Figure 2. A mutant strain, blocked so that it cannot decarboxylate tris(hydroxymethyl)acetate, accumulates this acid when grown on an assimilable carbon source, such as acetate, in the presence of pentaerythritol. Pentaerythritol, originally the carbon source for the wild type strain, becomes the starting material for bioconversion by the mutant. Of considerable interest is that this mutant is able to perform this type reaction (oxidation of a primary alcohol to a carboxylic acid) on a variety of analogs. Blocks at other points on the metabolic pathway will result in the accumulation of metabolites preceding the block. This example illustrates some of the potential of the enrichment method in bioconversion investigations.

A typical enrichment procedure is exemplified by the isolation of aerobic chemoheterotrophs that grow on the water soluble vitamin, pantothenic acid, as the only source of carbon and nitrogen.[18] The isolation medium contained per liter 0.27 g KH_2PO_4, 0.26 g $MgSO_4 \cdot 7H_2O$, 2.8 mg $FeSO_4 \cdot 7H_2O$, 1.5 mg $MnSO_4$, 2.1 mg Na_2MoO_4, and 2.57 g potassium pantothenate. The salts medium was adjusted to pH 7.0 and autoclaved separately. After cooling the pantothenate, sterilized by filtration, was added. The primary enrichment was conducted in a 250-mℓ Erlenmeyer flask containing 125 mℓ of the medium and an inoculum consisting of rich mud gathered from water at the edge of a stream. The amount of mud in the flask was adjusted so that the depth of the liquid over it was about 2 cm. This was done to promote aerobic conditions in the medium. The flask was incubated undisturbed at 30° for about 6 days. An initial spurt of growth subsided and was followed by a gradual increase of turbidity during the incubation period. Nonspecific organisms growing on the nutrients extracted from the soil accounted for the initial growth; what followed was the gradual predominance of pantothenate utilizers. When the culture became very turbid, 0.1 mℓ was used to inoculate fresh pantothenate medium in flasks. These flasks were incubated with shaking for about 24 hr, and the transfer process was repeated. This was done several times over and then the culture was plated out onto pantothenate agar. Several types of colonies were found that were able to degrade pantothenate.

This method can be used to isolate a variety of microoganisms with the ability to metabolize many different organic carbon and nitrogen compounds. The timing factor is very important in enrichment isolation procedures. In the primary enrichment vessel there will be a progression of organisms that develop from the complex population in the inoculum, and then lose out to successors as they adapt to changes in the medium wrought by predecessor metabolic activities. The desired strains may be lost, if the culture is transferred to fresh medium, or plated out at the wrong time. The way to avoid this loss is, simply, to make one's transfers at several different times as the incubations progress. Additional discussion of the enrichment procedures can be found in articles by Aaronson,[19] Veldkamp,[20] and Beck.[21]

V. BIOCONVERSION PROCEDURES

A. Outline of a Generic Procedure

1. Prepare culture system. A selected culture is grown to a desired state. The resultant cell mass in sundry ways becomes the catalyst for the bioconversion reaction. action.
2. Add substrate. Both timing and method of addition of substrate are important factors to consider.
3. Add effectors. Effectors are compounds that modify enzyme reactions. Inhibitors and activators are examples of effectors. Side reactions can be subdued by the effective use of selective inhibitors. Timing and method of addition of these compounds also are important.
4. Incubate the reaction mixture. The reaction is incubated at acceptable, but not necessarily maximal, values for the environmental parameters, until the transformation is complete. Optimization comes later.
5. Monitor reaction. Samples are taken at well-timed intervals and monitored for levels of products, starting material, pH value, cell mass, energy source, and any other useful parameters.
6. Terminate reaction. When sample ananlysis indicates that product accumulation has ceased, the reaction is stopped by physical removal of active cell mass.
7. Product isolation. At this point the bioconversion process becomes a separation and purification in organic chemistry. The main characteristic of bioconversions are that the solvent is mostly water, and the products are in relatively low concentration.

It is seen that bioconversion reactions essentially are catalytic chemical reactions, the main point of departure being in the methodology of preparing the catalyst, and the art of microbiology required to do this.

B. Types of Culture Systems

There are numerous ways to use cells in bioconversion processes. Each has its mix of advantages and disadvantages.

1. Batch Culture

Many biotransformation studies begin in batch cultures, for this is a direct method, and it is convenient to use for screening. In unadorned simplicity, this method involves the preparation and inoculation of a medium, addition of substrate, and incubation until completion of the reaction. The timing of substrate addition is discussed below. When quantitative studies are underway, it is advisable to use a calibrated inoculum.

Mid log phase to late log phase cells are the best for inoculum use. Both the state of growth and the cell density should be standardized. The number of cells in the inoculum should be between $1/10^4$ to $1/10^2$ of the number of cells in the fully grown culture. If the inoculation medium is different than the growth medium, it may be desirable to collect and wash the cells, and resuspend them in water or saline. In this way unwanted additions to the bioconversion medium are avoided.

2. Continuous Culture

In continuous culture, cells are maintained in a steady state of growth for relatively long periods by continuous addition of fresh medium, and withdrawal of spent medium at the same rates. In the chemostat method, the medium is constructed so that a nutrient, such as the carbon or nitrogen source, is limiting to growth. The growth rate, therefore, is controlled by the addition of fresh medium. Detailed explanation of this system can be found in *Methods in Microbiology.*[22] Continuous culture systems so far have not found much success in bioconversions, but they might have potential to be realized in the future.

3. Resting Cells

Resting cells are live, nongrowing cells that retain many of their enzyme activities. Usually they are removed from their spent medium, and resuspended in an incomplete medium (e.g., one without a nitrogen source), so that growth is not possible. When these cells are exposed to starting materials and incubated in an appropriate physical environment, bioconversion will occur. Conditions for incubation can be used that facilitate the bioconversion even though they are inimical to growth. High cell densities can be used to speed up the process.

4. Dried Cells

Dried cells are another version of resting cells. Microbial cells will survive, or at least many of their enzyme activities will survive, certain drying methods. Two of the most useful methods are lyophilization and acetone treatment. Lyophilization is just a larger scale application of the drying method previously described for the preservation of cultures. Cells washed by centrifugation or filtration are resuspended in dilute buffer or water. The suspension in a round bottom flask is frozen in a thin shell (preferably no more than 5 mm thick) with a dry ice-acetone bath. When the flask is placed under a high vacuum, water is removed and a fluffy powder is obtained. This powder can retain its activity for years, if it is stored frozen.

Acetone powders are prepared by treating a thick cell suspension with cold (−20°) acetone at least three times. After each slurrying in acetone, the cells are collected by filtration with suction. In the final stage a wash with cold ether helps to remove residual acetone. Acetone powders also should be stored frozen. The methods of using dried cell powders are the same as those used for fresh resting cells, but the dried preparations have the convenience of ''instant powders''. An additional convenience is that aseptic conditions need not be maintained for biotransformations with resting cells, including the fresh ones.

5. Spores

Fungi that produce conidiospores or sporangiospores offer an important and expediant alternative for biotransformations. Bacterial endospores generally are inactive, but many of the fungal spores have useful enzymatic activities. See the review by Vezina and Singh for an excellent discussion of this topic.[23] Spores suspended in a medium that does not allow germination can be used as a relatively stable source for biotransformation activities. Moreover, as in the case with resting cells, aseptic conditions do not have to be maintained in these processes. Fungi vary in their capacity to produce useful quantities of spores. It is not always possible to find the conditions for useful sporulation. A medium that works well for *Aspergillus ochraceus* is moistened wheat bran under conditions of high humidity.[23] Spores are harvested from surface cultures by washing with a dilute surfactant, such as Tween 80 at 0.01%. During the washing the surface is gently teased with a sterile rod or needle. Mycelial fragments can be removed by filtration through sterile glass wool; the spores can then be collected by centrifugation. The resulting paste should be stored at −20°. Some types of spores are stable for years frozen. Others are stable for only a few months; still others must be used immediately, since they are not stable frozen or otherwise. Liquid nitrogen storage also has been used with success.

Once obtained, the spore suspensions are used for bioconversions similiarly to the ways that resting cells are used. The medium should not be complete in the major nutrients, but it should contain a buffer and generally an energy source, such as glucose. To such a medium the substrate is added, and the incubation is conducted at the proper conditions of temperature and agitation. The usual concentration of spores in an incubation is 1×10^8 to 1×10^9 spores per mℓ. At the lower pH values of 5 and 6, bacterial growth is discouraged. Antibiotics also can be added if bacterial contamination is a problem. Spores usually do not germinate in these incubations.

6. Permealized Cells

Techniques are available for altering the permeability of cells, such that the entry of substrates and the exit of products is facilitated. These techniques involve treatments with surfactants[24] or solvents.[25] Since the alteration of permeability can drastically affect cell viability, the application of these procedures mainly are practical to use with resting cell preparations. However, there are some growing cell processes that advantgeously employ permeability altering agents. An example is found in the production of glutamic acid.[26] A clever application is the use of antibiotics, such as penicillin, at levels that affect cell wall synthesis enough to enhance permeability, but are not enough to kill the cell.[27]

7. Immobilized Cells

Immobilized cells are cells that either have been fixed or bonded to a surface, or have been entrapped in a gel matrix. Both methods of immobilization have achieved a measure of success in the laboratory and the factory.[28] Immobilized cells retain their activities over surprisingly long periods, in the range of months. Compared to unsupported resting cells, significant advantages are longer catalyst lifetime, the possibility of construction of a continuous system, and the possibility of the regeneration of cells *in situ* by irrigation with a nutrient medium. The investigation of immobilized systems resolves into two phases: the immobilization processes themselves, and the design of a workable system for their use. There are many published methods for cell immobilization and use.[29-32] Two immobilization methods that have rather broad practical value are polymerization entrapment in polyacrylamide, and physical entrapment in carrageenan.

Polyacrylamide method — The following method of Chibata and co-workers was optimized for entrapment of *Escherichia coli* cells.[32] To a suspension of cells (10 g wet weight per 40 mℓ physiological saline) were added 7.5 g acrylamide, 0.4 g *N,N*-methylenebisacrylamide, 5 mℓ of 5% β-dimethylaminopropionitrile, and 5 mℓ of 2.5% potassium persulfate. After standing for 10 to 15 min at a temperature not exceeding 40°, the stiff gel was cut into 2 to 3 mm cubes. The entrapped cells were washed thoroughly before use for enzyme reactions.

Carrageenan method[32] — *Escherichia coli* cells (5 g) having aspartase activity were

Table 6
LIST OF USEFUL MICROBIOLOGICAL EQUIPMENT FOR A BIOCONVERSION LABORATORY

Item	Specifications[a]
1. Pipets	Disposable, glass, serological, cotton plugged, sterile, 1,5,10-mℓ.
2. Petri dishes	Disposable, optically clear, polystyrene, 100 × 15 mm.
3. Culture tubes	Disposable, borosilicate glass, 18 × 150 mm.
4. Culture tube closures	Test tube caps, polypropylene, disposable, 18 mm.
5. Filter discs	6.5 in. diameter, type A400-33, Great Lakes Filter, Detroit, Mich.[b]
6. Autoclaving bags	Polypropylene bags, 10 × 15 in.
7. Inoculating loops	24 gauge nichrome wire, 3 mm loop, about 70 mm long.
8. Inoculating needles	24 gauge nichrome wire, about 50 mm long.
9. Loop and needle holder	Needle holder after Rosenberger and Greenman or similar type.
10. Autoclave	Sterilizer, steam-powered, 20 × 20 × 36 in chamber size.
11. Microbiological hood	Vertical laminar air flow, HEPA filter.
12. Shaker[c,d]	Orbital, 0-400 rpm, 2-in. throw.
13. Incubator[c]	Many possibilities. Need at least 20 to 60° capability.
14. Anaerobic jars	Vented lid for evacuation and exchange of gases, catalyst system for removing oxygen.
15. Liquid nitrogen system for culture storage	Check laboratory instrument guide.
16. Lyophilization system	A variety of units are available from laboratory supply houses.
17. Cell breaking devices	
a. Probe type sonic	10 to 20 kilocycles needed.
b. Bead mill	Various types are available.[e]
18. Centrifuges	High speed, refrigerated.
19. Screw cap culture tubes	Tubes, 20 × 150 mm with plastic screw caps with rubber liners.
20. Sterile filter units	Disposable, sterile membrane filter assembly, 0.20 μ porosity.
21. Microscope	Bright field and phase contrast optics, eyepieces, 10×, objectives, 10×, 40×, 100×.
22. Culture tube basket (for storing slants)	Rectangular, tilt top.
23. Test tube cabinet (for incubating slants)	Stainless steel with racks for 18 × 150 mm tubes.

[a] Most of these items can be found in general laboratory supply catalogs.
[b] Other types are available.
[c] Combined incubators and shakers are recommended.
[d] A controlled temperature facility is essential. A temperature range of 5 to 65° is desirable.
[e] See Nickerson, W. J., *Handbook of Microbiology,* Vol 2, Laskin, A. I. and Lechevalier, H. A., Eds., CRC Press, Boca Raton, Fl., 1973, 843.

suspended in 5.0 mℓ of 0.9% saline at 45 to 50°; carrageenan (1.7 g) is dissolved in 34 mℓ saline at 45 to 60°. The two fluids were mixed, and then cooled to about 10°. The gel strength was increased by soaking in cold 0.3 *M* potassium chloride. This gave in a stiff gel which could be granulated. Treatment with hardening agents such as glutaraldehyde and hexamethylenediamine produces a more stable gel. The activity remaining in the preparation after this treatment was about 33%.

8. Broken Cells and Enzyme Extracts

The ultimate in permealized cells are broken cells. These are cells whose walls have been fractured. Interestingly enough, most cells of microorganisms are not easily broken. The bacterial cells as a rule are the toughest. The breaking methods with the widest application are grinding and treatment with high intensity sound waves. Sound wave treatment is termed sonication. Probe sonifyers are available from several manufacturers (See Table 6). The equipment is easy to use, and is especially good for the disruption of bacterial cells, but has use for yeast and fungi also. The method involves suspending cells at 30 to 40 g wet weight per 100 mℓ of buffer, and the application of sound energy through a probe immersed in the suspension. Usually short bursts at maximum intensity are the most effective, because the time between can be used for dissipation of heat. During the treatment the cell suspension is cooled in an ice bath. Microscopic investigation will give an indication of how effectively the cells were broken.

For use in bioconversion methods, the substrate is mixed with suitable portions of the broken cells. Depending on the type of enzymatic activity being utilized, it may be necessary at this stage to add enzyme cofactors; e.g., for oxidation-reduction with dehydrogenase enzymes it will be necessary to add the appropriate pyridine nucleotide cofactor. Enzymes in broken cells have shorter half-lives than their counterparts in intact cells, but activities can be preserved for moderate lengths of time by freezing.

Various gadgets for grinding cells are listed in Table 6. These should be especially useful for breaking the fungal and yeast cells.

Cell extracts are obtained from broken cell suspensions by centrifugation to remove insoluble debris. Extracts and broken cell suspensions have about the same uses in bioconversions. However, obtaining an extract is the prelude to enzyme purification, the discussion of which is beyond the scope of this chapter.

C. Substrate Addition

This section is a discussion of the methods of addition of starting materials to a bioconversion process, although many of these methods are just as applicable to the addition of carbon sources (or other compounds) to a culture medium. The main difficulties are encountered when water-insoluble materials must be added to aqueous systems. The simplest instances occur when the substrate is water-soluble. For these compounds the questions that must be answered are how much to add, and when to add it. The amount to add depends upon the type of compound, and also how much product one wants to accumulate from the process. Of course, increasing the amount of substrate in relation to the amount of active cell mass will increase the time required to complete the bioconversion reaction. The maximum amount of substrate that can be added depends on the nature of the compound to a large extent. Uncharged compounds can be added in rather large amounts; an approximate upper limit would be about 10% (w/v), although higher concentrations may be possible. For ionic materials the maximal levels, as a rule of thumb, should be about half this (about 5%). These are the amounts suggested for use with nonosmophilic microbial cells.

Another matter that must be considered is the relative toxicity of the substrate; this may lower considerably the upper limit of tolerable concentration. The time of substrate addition is also a matter to be considered. Compounds are more likely to exert adverse effects when they are added at the beginning of the growth phase. Such effects are least of all in resting cell preparations of the various types. The addition of a compound in increments is a useful technique for the avoidance of toxic concentrations. When working with higher concentrations of materials, it is well to remember that the accumulation of large amounts of even a relatively nontoxic product also can be inhibitory.

The addition of water-insoluble materials presents a different set of problems. There are many instances where satisfactory results are obtained when insoluble solids are added as fine powders. This is the method frequently used in steroid conversions.[33] To be sure, the addition of powders hardly affects the osmotic pressure of the medium, which eliminates one worry, but the fundamental problem of achieving the best contact possible between substrate and cells remains. The addition of neat liquids seems to work fairly well; for example, the addition of α-tocopherol acetate to a salts medium allows luxuriant and rapid growth of bacteria able to metabolize this compound.[34] Other examples of this phenomenon are drawn from studies of hydrocarbon biotransformations.[35] Even gases such as methane, when bubbled into a medium, are effective substrates.

The addition of insoluble substrates as solutions in water-miscible organic solvents is a venerable method. Favorite solvents are the lower alcohols, acetone, dimethylformamide, and dimethylsulfoxide. The latter compound, while relatively nontoxic, has an unpleasant odor, especially when it is reduced to sulfides in the culture medium. (However, a little odor never stopped a synthetic organic chemist.) The usual way of employing these solvents is to make a very concentrated solution of the substrate so that the amount of solvent added to the bioconversion medium is as little as possible. Before a solvent is used, it is wise to determine the tolerance of this solvent by the culture system.

Surfactants also are used to disperse insoluble materials. There are several nontoxic surfactants that are available; the Tweens' and Brijs' are examples. If the surfactant is intended for use with nongrowing cells, either in the stationary phase or in some form of the resting state, a toxic surfactant can be used (e.g., Triton X-10C). Substrate can be added directly as a liquid or powder to a growth or incubation medium containing up to 3% surfactant. Substrates also can be added as dispersions or even solutions in surfactants. Combinations of surfactants and water-miscible solvents may offer some advantages.

The time of addition of substrates to incubation mixtures becomes crucial when a process with growing cells is involved. Fortunately, the addition of even small amounts of substrate at the inception of growth frequently induces desired enzymes, thereby improving the biotransformation process. In other cases, early addition clearly would be detrimental. The optimal times must be determined by experiment.

D. Effectors and Precursors

Effectors are compounds that modify enzyme activity. The term perhaps is more properly used in enzymology than it is in the microbial transformation of compounds. Nevertheless, ultimately it is the enzymes in the cell that are responsible for the chemical reactions that interest us. When dealing with whole cells, there is a layer of complication over the ordinarily complicated phenomena of enzymology. Two of these additional complications are cell permeability, and the occurrence of enzyme induction. Accordingly, we shall borrow the word "effector" and use it to refer to compounds that affect enzymatic activities of the cell. Activities can be affected two ways. Already existing activity can be modified by inhibition or activation, and the amount of activity can be increased or decreased.

Inducers — The amount of enzyme activity in a cell suspension is a function of the amount of enzyme per cell multiplied by the number of cells. Obviously, activity can be increased by increasing the number of cells. The way to increase the amount of activity per individual cell is to exercise the processes of enzyme induction (or derepression). The synthesis of enzymes is a precisely controlled process. Some enzymes are not made in detectable amounts unless an inducer is present. These are the inducible enzymes. Constitutive enzymes are present under the general circumstances of growth, and do not increase in the presence of inducers. The early stages of logarithmic growth are the best times to add inducers. Frequently the apparent inducer of an enzyme is also its substrate. This is convenient. Gratuitous inducers are not substrates for the enzymes they induce. Full induction of an enzyme can be obtained with a small amount of a gratuitous inducer, because the enzyme cannot destroy it, as it does its own substrate. Isopropyl β-thiogalactoside is an effective gratuitous inducer of β-galactosidase in *Escherichia coli*.[36]

Inhibitors — Another important type of effector is the selective inhibitor. The objective of the use of inhibitors is to inhibit unwanted side reactions, while having a minimal effect on the main bioconversion activity. An example of the current use of inhibitors is the selective degradation of the side chain of plant and animal sterols to obtain androstenedione. It has been observed that enzymes degrading the nuclear structure of the sterols require divalent iron. If a powerful chelator of divalent iron is added to the bioconversion medium in proper amounts, the total degradation of the steroid is stopped at the point where the side chain at ^{17}C is clipped off with the formation of an oxo group.[37]

Precursors — The addition of precursors has an important place in the history of the development of penicillin biosynthesis on an industrial scale. The addition of phenyl acetic acid was found to greatly stimulate production.[38] The addition of benzaldehyde to "active aldehyde" fermentations produces an acetoin type compound that is converted chemically to D-ephedrine.[39] The technique of mutasynthesis is based on the addition of precursor analog to a mutant strain unable to synthesize its own precursor. The added analog finds its way into the biosynthetic sequence, resulting in the synthesis of end-product analogs. This technique is illustrated in synthesis of streptomycin compounds.[40]

E. Liquid Culture Incubation Methods

For bioconversion studies it is essential to have incubation methods available that are convenient, and that allow for relatively large numbers of individual experiments. Three factors essential to liquid culture incubation are mixing, aeration, and temperature control. Temperature control is achieved easily with available commercial equipment, and so it will not be discussed. Although many different ways have been developed for the liquid culture of microorganisms, there is no single perfect system. However, the one system that has the most to recommend itself for initial studies and screening, as well as for preliminary scale-up and yield improvement studies, is the shaken flask. The standard conical Erlenmeyer flask shaken on a rotary shaker in a controlled environment provides the best all-around liquid culture system for small scale work. Suitable flask sizes are from 125 to 2000 mℓ. Recommended medium volume is 1/10 of flask volume to get the best aeration. We suggest standardization of both flask and shakers in a laboratory. Most of our work is done in 250-mℓ flasks on rotary shakers with a 2-in. throw. If larger volumes for sampling are required, as in fungal cultures, the larger 500-mℓ flasks are used. For preliminary scale-up to liter volumes, we suggest the 2.8 ℓ wide-mouthed Fernbach flask with 500 to 1000 mℓ working volume. The flask should be baffled to provide the turbulance that enhances aeration efficiency. All flasks are covered with milk filter discs. Table 7 shows the effect

Table 7
EXAMPLE OF OXYGEN TRANSFER IN SHAKEN FLASKS[a,b]

	$K_L \cdot a$[c]	
Volume in flask (mℓ)	**Shaker at 125 rpm**	**Shaker at 200 rpm**
25	1.86	2.19
50	1.81	—
75	1.03	1.64
100	0.72	1.41
125	0.78	1.06

[a] The experiments were conducted at 30° in 250-mℓ conical flasks on a rotary shaker with a 2-in. throw. The flasks were covered with milk filter discs.

[b] Personal communication from P. Masurekar. The procedure was based on the method of Hsieh et al.,[42] which involved the oxygen-linked oxidation of glucose to gluconic acid via glucose oxidase.

[c] Volumetric mass transfer coefficient. This is a measure of oxygen transfer efficiency; higher values indicate higher efficiency.

of culture medium volume and shaking speed on aeration efficiency in small flasks.[41] The addition of 0.005 to 0.1% Polyglycol P-2000 (Dow Chemical Co., Midland, MI) to media in the larger flasks controls foaming. The glycol can be sterilized with the medium.

Another way to achieve mixing in liquid culturing is to use stirring. The magnetic system with a stirring bar is the easiest one to use because the maintenance of sterility is not a problem. This type of stirring is useful for maintaining a suspension of cells in instances where aeration by turbulence is not desired, such as a yeast process that is essentially anaerobic, or any other anaerobic process.

The best vessel easily obtainable for conducting bioconversions is, of course the stirred fermentor. This applies especially to aerobic processes. However, anaerobic processes run very well in these fermentors under an atmosphere of nitrogen. A good laboratory fermentor represents a considerable investment, but perhaps an essential one if the bioconversion processes are not to remain mere laboratory curiosities. Whether the process involves growing cells or resting cells, the fermentor has a key place, and that is the efficient production of the catalytically active cell mass. The aeration attainable in a well-designed fermentor is much superior to that obtained with shaken flasks. A certain amount of scale-up can be accomplished with the larger flasks, but for eventual process development, work in stirred fermentors is essential. The operation of laboratory fermentors has been discussed by Calam[43] and Davis and Blevins.[44]

Immobilized systems (especially immobilized cells in some sort of gel) can be incubated and tested for activity in shaken flasks. This accomplishes the mixing and aeration (if needed) at the same time, and it can be done gently enough to avoid damage to the gel. More sophisticated would be a packed column system for continuous operation. A glass column can work efficiently, if it is not necessary to provide oxygen. The provision of adequate oxygen is difficult, although a well-sparged substrate solution will help matters. A method for injecting air at positions along the axis of the column would be preferred. Another solution is to use the fluidized bed technique in which air enters at the bottom of the column. For improved aeration performance, spargers also could be placed along the vertical axis of the column.

F. Monitoring

It is essential that bioconversion reactions be monitored, even in cases where the process is known very well. One must know what is happening in the reaction mixture to conduct it properly: for example, if a reaction goes too long it may use up all the starting material, as well as all the product, or side reactions may generate proucts that make purification difficult. Both the appearance of product and the disappearance of starting material should be followed. Monitor the pH value, and any other parameter for which a reasonably convenient measurement method exists. The size of the sample taken is determined by the concentration of reactants and the sensitivity of the analytical method. It also is determined somewhat by the type of microorganism used in the process. A culture of well-dispersed cells with abundant growth will need relatively small samples to provide representative data. Organisms that grow in nonhomogeneous masses will cause difficulty in sampling. The matter is even more complicated when the substrate is insoluble, for frequently the insoluble compound is associated with the mass of growth, making the culture anything but homogeneous. There are several ways to handle this problem. First, attempts should be made to find a culturing method that gives a more homogeneous growth pattern. This sometimes is possible with fungi, if one insures that the inoculum consists of well-dispersed mycelial fragments. Inoculation with a spore suspension should give a reasonably uniform culture. Changes in medium constituents also may help. If larger samples must be taken, the volume of the culture fluid must be large enough to accomodate the loss due to sampling. For bacterial cultures the sample size can be about 1 mℓ. For the mycelial cultures, 5 to 10 mℓ is suggested.

In a growth-associated process the cell mass should be followed with time. If culturing must be done in small flasks, one way to side-step the sampling problem is to set up parallel flasks so that an entire flask per time point can be used as the sample. Sampling should be done at shorter intervals as the reaction approaches completion. Usually it is good practice to do a pilot experiment do get some idea of the overall time frame of the reaction. Another good practice is to run duplicate flasks in the experimental runs. Anything that can be measured within reason should be monitored. For instance, if ammonia is given off or taken up in the reaction, it should be measured. It would be useful to measure the dissolved oxygen levels, but this is not done very well in flasks. This is an experiment that must wait for the fermentor runs.

G. Product Isolation

We will discuss here only those matters that are rather different from the usual chemical reaction work-up. The main common point in all biotransformations is that they are done in water solution. Therefore, the products must be retrieved from this environment. The problem is almost trivial, if the product can be extracted from either an acidic or basic solution. Ion exchange resins are quite useful, especially for isolating amphoteric compounds.

More often than not the product is found in the culture fluid after the removal of the cell mass. Cell mass is removed by filtration, or by centrifugation. The reaction is stopped merely by the removal of the biocatalyst. If, however, the reaction cannot be worked up immediately, it can be preserved by cooling in a refrigerator. Freezing is allowable, if the thawing process does not cause cell breakage which releases significant amounts of extraneous material, thereby unnecessarily complicating the work-up. If the product is found in the culture supernatant, this is fortunate. If not, the cells must be extracted. Usually this circumstance occurs only when insoluble starting materials and products are encountered. Even so, it is surprising that many insoluble products are dispersed in the culture fluid and not adsorbed to the cells. When this mass is extracted with an organic solvent, considerable extraneous material is solubilized, and the isolation of the desired product is more difficult. Of course, reaction systems based on immobilized cells are practical only when the product is released to the medium.

H. Yield Improvement

Once an interesting reaction is discovered, one objective of further work is to obtain enough of the product for its identification, and second, to make enough so that it can be submitted for testing. In order to do this, yield improvement and preliminary scale-up studies are required. Yield improvement can be accomplished by the optimization of nutritional and environmental factors. There are statistical methods for optimization that are effective.[45] However, for this chapter we shall discuss an optimization method that consistently yields good results in our laboratory. The general approach of the method is to vary one parameter at a time; then, incorporate its discerned optimum in the next round of tests with a new parameter, always keeping a common experiment from the last experiment to the next one. Finally, a process incorporating all the optimum conditions that have been built one on the other can then be tested separately in each of its parameters all over again. However, the degree of yield improvement in the second round will be less than in the first round.

1. Physical Environment

Temperature — The first parameter to test for effect on yields of product is the temperature. Do each variation point in duplicate flasks. If the results are widely separated, repeat the experiment. (This caution applies to all the experiments with the various parameters.) At first, probe no more than 5° on either side of the optimal growth temperature of the particular species of microorganism. If the effects observed are insignificant, extend the range to 10° on either side. A change of only a few degrees could make a large difference in yields. This was the case in the production of L(+)-3-hydroxyisobutyrate by *Pseudomonas putida*.[46] If the bioconversion is conducted with resting cells or stationary phase cells, the effect of temperature should be studied for both the growth phase and the production phase.

Aeration — For aerobic processes in shaken flasks, this parameter can be studied up to the point of aeration limitation. These studies should be done at the optimal temperature found in the previous experiments. Study both the production and the growth phase, if they are different. Conclusive aeration studies must be performed in a well-equipped laboratory fermentor, with which it is possible to obtain higher mass transfer rates. Also, dissolved oxygen and exhaust gas composition data are much accumulate with an instrumented fermentor.

pH Value — This parameter is difficult to study in flasks where the continuous monitoring and controlling of pH are not practical. The effects of pH are best left for the fermentor investigations. However, if the optimal pH value for growth of the microorganism in question is not published, it can be investigated by varying the starting pH of the growth medium. If the value changes during growth or bioconversion stage, it can be noted, and a different starting value can be tested for effect on the rate and extent of growth. If the medium becomes acidic during growth, the addition of finely ground calcium carbonate (0.5 to 1.0%) controls the pH at about 6.8.

2. Growth Medium

While incorporating the optimal environmental conditions determined previously, the components of the growth medium are varied and tested for effect one at a time.

It should be noted that for each change of condition the growth kinetics and bioconversion kinetics will change. Therefore, sampling times should be adjusted to account for these changes. Variations in the carbon source type and concentration should be investigated first. Frequently, in media for chemoheterotrophs, the carbon and energy sources are the same. Of course, if they are different, they should be varied independently. A series of carbon sources should be tested at the same concentration as the original carbon source. For example, if the original source is glucose, other monosaccharides, including sugar alcohols and glycerol should be checked. Also, try citric acid cycle intermediates, as well as pyruvate and acetate, for their effects. Combinations can be superior to a single carbon source. Furthermore, try an array of auxilliary carbon sources, especially if the biotransformation substrate and the main carbon source are identical. When the better carbon sources are found, the effect of their concentrations in the medium should be checked. Changes in the carbon source can have marked effects on enzyme induction. Glucose will suppress the induction of some enzymes. Changing to glycerol may correct the problem.

It is advisable to investigate media components in the order of their concentration. The next most abundant component is the nitrogen source, which should be checked for effect of changes in type and concentration. The simple organic salts, ammonium sulfate, ammonium nitrate, and potassium nitrate, the amino acids, glutamate and asparagine, and urea are among the first to be looked at. Next investigate the more complex organic sources: the yeast extracts, peptones, tryptones, and other digests, such as corn steep liquor. Also try combinations.

Sulfur and phosphrus sources, trace element, and growth factor sources can be investigated next, but these substances are present in such low concentrations in the medium that it is necessary to return to a basal, chemically defined medium in order to conduct meaningful experiments. If by this time a significant improvement in yields has not been achieved, it may be time to start the screening process for better strains. Or, it may be time to consider mutational procedures for strain improvement.

3. Mutation

The procedures available for strain improvement by mutation and other genetic manipulations have developed extensively over the last 35 years. The topic is too large for the scope of the present chapter. Certainly, some of the techniques, such as mutagenesis by treatment with ultraviolet radiation or chemical mutagenesis with nitrosoguanidine are relatively simple in concept. However, the production of mutants is only a part, perhaps a minor part of, the task; the difficulties are in the selection of the mutants for desired characteristics, such as improved productivity or metabolic blocks. The reader is refered to the excellent writings of Hopwood (see Hopwood and Merrick[47] for a recent review) and Elander and Chang[48] for background material and discussions of techniques.

VI. A CULTURE COLLECTION FOR THE NOVICE

After the newcomer to the area of bioconversions of organic compounds has equipped the laboratory and grappled with the necessary microbiological techniques and concepts, he is still faced with the task of accumulating a collection of microorganisms. Indeed, the culture collection is the basic tool box for the investigator. To be sure, cultures will be collected from the various sources as mentioned earlier in this chapter, but to make the collection process a little easier, I have put together a starter set of cultures selected from the molds, yeasts, and bacteria that have been used for biotransformations. It is hoped that this group of cultures will provide a foundation upon which to systematically build a valuable and functional collection. Although all

Table 8
MICROORGANISMS FOR A BEGINNING CULTURE COLLECTION

Bacteria	ATCC No.	Medium[a]	Temp.	Comments
Eubacteriales				
Acinetobacter calcoaceticus	14987	1	26°	Aerobic
Alcaligenes eutrophus	17697	1	30°	Aerobic
Arthrobacter simplex	6948	1	26°	Aerobic
Bacillus subtilis	82	1	30°	Aerobic
Clostridium acetobutylicum	4259	2	37°	Anaerobic
Escherichia coli	9637	1	37°	Facultative
Flavobacterium heparinum	13125	1	30°	Aerobic
Lactobacillus leichmannii	7830	3	37°	Facultative
Micrococcus roseus	416	1	26°	Aerobic
Streptococcus faecium	8043	3	37°	Facultative
Pseudomonadales				
Gluconobacter oxydans	621	4	25°	Aerobic
Pseudomonas putida	21244	1	26°	Aerobic
Actinomycetales				
Corynebacterium species	21245	5	30°	Aerobic
Mycobacterium rhodochrous	19067	5	37°	Aerobic
Nocardia corallina	14347	5	26°	Aerobic
Streptomyces argenteolus	11009	5	26°	Aerobic
		Yeasts		
Ascomycetes				
Debaromyces polymorphus	20280			All cultures will grow on medium 7
Hansenula anomala	20170			
Kluyveromyces lactis	2628			All cultures will grow at 24° - 26°
Lipomyces lipofer	10742			
Pichia membranaefaciens	2254			
Saccharomyces cerevisiae	15248			
Basidiomycetes				
Bullera alba	18568			
Ittersonilia pyriformis	15496			
Sporobolomyces pararoseus	11386			
Tilletiopsis albescens	24343			
Fungi imperfecti				
Brettanomyces petrophilum	20224			
Candida tropicalis	650			
Kloeckera africana	20111			
Rhodotorula glutinis	15125			
Torulopsis magnoliae	12573			
		Fungi		
Phycomycetes				
Cunninghamella echinulata	36190	6		All cultures grow at 24° - 26°
Mortierella zonata	13309	6		
Rhizopus stolonifer	10404	6		
Thamnostylum piriforme	8992	6		
Ascomycetes				
Cordyceps militaris	34164	6		
Eupenicillium javanicum	26879	7		
Neurospora crassa	10815	8		
Thielavia terricola	13807	6		
Basidiomycetes				
Boletus luteus	36646	7		
Coniophora puteana	12675	7		
Lentinus lepideus	36557	6		

Table 8 (continued)
MICROORGANISMS FOR A BEGINNING CULTURE COLLECTION

Bacteria	ATCC No.	Medium[a]	Temp.	Comments
Fungi imperfecti				
Aspergillus niger	9142	6		
Beauveria bassiana	7159	9		
Curvularia lunata	12017	6		
Septomyxa affinis	6737	6		

[a] See Table 5 for composition of media.

of the strains listed are available from the American Type Culture Collection, many, if not all, can be found in the other major collections around the world. I have left out microorganisms that require a U.S. Department of Agriculture permit for transport (also known pathogens), and in so doing we have neglected some strains with illustrious histories in biotransformation investigations, for example *Calonectria decora*. Such organisms can be obtained later, as needed, when the researcher is experienced in microbiology. The suggested starter collection is shown in Table 8. The table includes the ATCC number, and suitable growth temperatures and media, as well as some comments.

VII. CONCLUDING REMARKS

The biotransformation of organic compounds is a multi-disciplinary subject, drawing from organic chemistry, biochemistry, and microbiology. Much of the interest in this area comes from the organic chemist who perceives new applications to the problems of chemical synthesis. This chapter is intended primarily for those chemists and biochemists newly recruited to the study of microbial transformations. In consideration of this, it has been necessary to include sections on the rather fundamental techniques of microbiology, not to mention a brief description of the microorganisms used for biotransformations. There are numerous ways to manipulate microorganisms, to culture and preserve them, and to use them for biotransformations. I have attempted to give a guided tour, as it were, through the techniques of the microbiological aspects, providing enough information, it is hoped, for a sound beginning. Instead of attempting to be all-embracing in the discussion of methods, I have elected to describe mainly those techniques that have proven to be reliable for us over the years.

REFERENCES

1. **Stanier, R. Y., Douderoff, M., and Adelberg, E. A.,** *The Microbial World,* 3rd ed., Prentice-Hall, Englewood Cliffs, N.J., 1970, chap. 4.
2. **Booth, C., Ed.,** Methods in Microbiology, *Vol. 4,* Academic Press, New York, 1971.
3. **Buchanan, R. E., and Gibbons, N. E., Eds.,** *Bergey's Manual of Determinative Bacteriology,* 8th ed., Williams & Wilkins, Baltimore, 1974.
4. **Skerman, V. B. D.,** *A Guide to the Identification of the Genera of Bacteria,* 2nd ed., William & Wilkins, Baltimore, 1967.
5. **Skerman, V. B. D.,** Important culture collections, in *Handbook of Microbiology,* Vol. 1, Laskin, A. I. and Lechevalier, H. A., Eds., CRC Press, Boca Raton, Fla., 1973, 815.
6. **Kieslich, K.,** *Microbial Transformation of Non-steroid Cyclic Compounds,* Georg Thieme, Stuttgart, 1976, 4.

7. **Douglas, S. D.**, Microscopy, in *Manual of Clinical Microbiology,* 2nd ed., Lennette, E. H., Balows, A., Hausler, W. J., Jr., and Truant, J. P., Eds., American Society for Microbiology, Washington, D.C., 1980, chap. 2.
8. **Ainsworth, G. C., Sparrow, F. K., and Sussman, A. S., Eds.**, *The Fungi,* Vols. I - IV, Academic Press, New York, 1973.
9. **Stevens, R. B., Ed.**, *Mycology Guidebook,* University of Washington Press, Seattle, 1974.
10. *Difco Manual,* 9th ed., Difco Laboratories Inc., Detroit, 1953, 240.
11. **Perlman, D.**, Procedures useful in studying microbial transformations of organic compounds, in *Applications of Biochemical Systems in Organic Chemistry,* Part 1, Jones, J. B., Sih, C. J., and Perlman, D., Eds., John Wiley & Sons, New York, 1976, 53.
12. **McDaniel, L. E. and Bailey, E. G.**, Liquid nitrogen preservation of standard inoculum: gas-phase storage, *Appl. Microbiol.,* 16, 912, 1968.
13. **Calam, C. T.**, The evaluation of mycelial growth, in *Methods in Microbiology,* Vol. 1, Norris, J. R. and Ribbons, D. W., Eds., Academic Press, New York, 1969, chap. 16.
14. **Luria, S. E.**, *The Bacteria,* Vol. 1, Gunsalus, I. C. and Stanier, R. Y., Eds., Academic Press, New York, 1960, chap. 1.
15. **Vulcani, B.**, personal communication.
16. **Stanier, R. Y., Douderoff, M., and Adelberg, E. A.**, *The Microbial World,* 3rd ed., Prentice-Hall, Englewood Cliffs, N.J., 1970, 88.
17. **Goodhue, C. T. and Schaeffer, J. R.**, The synthesis of tris(hydrooxymethyl)acetic acid by fermentation of pentaerythritol, *Biotechnol. Bioeng.,* 11, 1173, 1969.
18. **Goodhue, C. T. and Snell, E. E.**, The bacterial degradation of pantothenic acid. I. Overall nature of the reaction., *Biochemistry,* 5, 393, 1966.
19. **Aaronson, S.**, Enrichment culture, in *Handbook of Microbiology,* Vol. 1, Laskin, A. I. and Lechevalier, H. A., Eds., CRC Press, Boca Raton, Fla., 1973, 725.
20. **Veldkamp, H.**, Enrichment cultures of procaryotic organisms, in *Methods in Microbiology,* Vol. 3A, Norris, J. N. and Ribbons, D. W., Eds., Academic Press, New York, 1970, 305.
21. **Beck, J. V.**, Enrichment culture and isolation techniques particularly for anaerobic bacteria, in *Methods in Enzymology,* Vol. 22, Jakoby, W. B., Ed., Academic Press, New York, 1971, 57.
22. **Tempest, D. W.**, The continuous cultivation of microorganisms, in *Methods in Microbiology,* Vol. 2, Norris, J. R. and Ribbons, D. W., Eds., Academic Press, New York, 1970, 259.
23. **Vezina, C. and Singh, K.**, Transformation of organic compounds by fungal spores, in *The Filamentous Fungi,* Vol. 1, Smith, J. E. and Berry, D. R., Eds., Edward Arnold, London, 1975, chap. 9.
24. **Martin, C. K. A. and Perlman, D.**, Stimulation by organic solvents and detergents of conversion of L-sorbose to L-sorbosone by *Gluconoacter melanogenus* IFO 3293, *Biotechnol. Bioeng.,* 17, 1473, 1972.
25. **Mizrahi, A. and Miller, G.**, Role of glycols and tweens in the production of ergot alkaloids by *Claviceps paspali, J. Bacteriol.,* 97, 1155, 1969.
26. **Aiba, S., Humphrey, A. E., and Millis, N. F.**, *Biochemical Engineering,* 2nd ed., Academic Press, New York, 1973, 81.
27. **Demain, A. L. and Birnbaum, J.**, Alteration of permeability for the release of metabolites from the microbial cell, *Curr. Top. Microbiol. Immunol.,* 46, 1, 1968.
28. **Tosa, T., Sato, T., Mori, T., and Chibata, I.**, Basic studies for continuous production of L-aspartic acid by immobilized *E. coli* cells., *Appl. Microbiol.,* 27, 886, 1974.
29. **Abbott, B. J.**, Immobilized cells, in *Annual Reports on Fermentation Processes,* Vol. 1, Perlman, D., Ed., Academic Press, New York, 1977, chap. 9.
30. **Abbot, B. J.**, Immobilized cells, in *Annual Reports on Fermentation Processes,* Vol. 2, Perlman, D., Ed., Academic Press New York, 1978, chap. 5.
31. **Kolot, F. B.**, Immobilized microbial systems: present state of development, in *Developments in Industrial Microbiology,* Vol. 21, Underkofler, L. A. and Wulf, M. L., Eds., Impressions Ltd., Gaithersburg, Md., 1980, 295.
32. **Chibata, I., Tosa, T., and Sato, T.**, Use of immobilized cell systems to prepare fine chemicals, in *Microbial Technology,* Vol. 2, 2nd ed., Peppler, H. J. and Perlman, D., Eds., Academic Press, New York, 1979, chap. 17.
33. **Perlman, D.**, Procedures useful in studying microbial transformations of organic compounds, in *Applications of Biochemical Systems in Organic Chemistry,* Jones, J. B., Sih, C. J., and Perlman, D., Eds., John Wiley & Sons, New York, 1976, 56.
34. **Goodhue, C. T.**, Specific growth of a soil microorganism on the natural isomer of α-tocopherol, *Biochemistry,* 4, 1822, 1965.
35. **Makula, B. A. and Finnerty, W. R.**, Microbial assimilation of hydrocarbons, I. Fatty acids derived from normal alkanes, *J. Bacteriol.,* 95, 2102, 1968.

36. **Jacob, F. and Monod, J.** Genetic regulatory mechanisms in the synthesis of proteins, *J. Mol. Biol.*, 3, 318, 1961.
37. **Martin, C. K. A.,** Microbial cleavage of sterol side chains, in *Advances in Applied Microbiology,* Vol. 22, Perlman, D., Ed., Academic Press, New York, 1977, 44.
38. **Behrans, O. K.,** Biosynthesis of penicillins, in *The Chemistry of Penicillin,* Clarke, H. T., Johnson, J. R., and Robinson, R., Eds., Princeton University Press, New Jersey, 1942, chap 19.
39. **Smith, P. F. and Hendlin, D.,** Further studies on phenylacetylcarbinol synthesis by yeast, *Appl. Microbiol.,* 2, 294, 1954.
40. **Shier, W. T., Rinehart, K. L., Jr., and Gottlieb, D.,** Preparation of four new antibiotics from a mutant of *Streptomyces fradiae, Proc. Nat. Acad. Sci. U.S.,* 63, 198, 1969.
41. **Masurekar, P.,** personal communication.
42. **Hsieh, D. P. H., Silver, R. S., and Mateles, R. I.,** Use of the glucose oxidase system to measure oxygen transfer rates. *Biotechnol. Bioeng.,* 11, 1, 1969.
43. **Calam, C. T.,** The culture of micro-organisms in liquid medium, in *Methods in Microbiology,* Vol. 1, Norris, J. R. and Ribbons, D. W., Eds., Academic Press, New York, 1969, chap. 8.
44. **Davis, N. D. and Blevins, W. T.,** Methods in laboratory fermentations, in *Microbial Technology,* Vol. 2, 2nd ed., Peppler, H. J. and Perlman, D., Eds., Academic Press, New York, 1979, chap 13.
45. **Campbell, R. C.,** *Statistics for Biologists,* Cambridge University Press, New York, 1974.
46. **Goodhue, C. T. and Schaeffer, J. R.,** Preparation of L(+)-β-hydroxyisobutyric acid by bacterial oxidation of isobutyric acid,., *Biotechnol. Bioeng.,* 13, 203, 1971.
47. **Hopwood, D. A. and Merrick, M. J.,** Genetics of antibiotic production, *Bacteriol. Rev.,* 41, 595, 1977.
48. **Elander, R. P. and Chang, L. T.,** Microbial culture selection, in *Microbial Technology,* Vol. 2, 2nd ed., Peppler, H. J. and Perlman, D., Eds., Academic Press, New York, 1979, chap 12.

Chapter 3

MICROBIAL TRANSFORMATIONS OF ANTIBIOTICS

Vincent P. Marshall and Paul F. Wiley

TABLE OF CONTENTS

I. INTRODUCTION

Microbial biotransformation of antibiotics has been an effective means for the preparation of analogs of members of several classes of antibiotics.[152,155,172-174,177,182] Although the scope of the conversions yielding clinically important antibiotics has been limited, these reactions have been of undisputed value in the synthesis of β-lactam antibiotics. Because of the specificity as well as the diversity of enzymatically catalyzed reactions, the potential remains to repeat the former successes with other antibiotic substrates. Recent studies of the bioconversion of the bleomycins suggest that such a potential may soon be realized for these antitumor antibiotics.

In this review the aim is to report as completely as possible all microbial transformations of antibiotics. The antibiotic classes covered include the aminocyclitol, macrolide, β-lactam, peptide, actinomycin, anthracycline, lincomycin, nucleoside, griseofulvin, chloramphenicol, novobiocin, and aureolic acid classes as well as other minor classes. This report is not limited to bioconversion by means of metabolic alteration of completely synthesized antibiotics, but additionally deals with such transformation means as directed biosynthesis and mutasynthesis.

Many of the compounds mentioned as substrates are not, strictly speaking, antibiotics, but because they are very closely related to antibiotics, they are included. In many cases which are included it was not shown by isolation that a new compound had been formed, but only that the substrate had been changed. These cases were decided on an individual basis, and some similar ones were excluded on the basis of personal judgment as to whether or not a transformation had been shown. The reference list is not exhaustive although complete coverage of antibiotic transformation was attempted. Some transformations were discussed in more than one publication and some of these publications were not included in the references.

II. BIOTRANSFORMATION OF ANTIBIOTICS

A. Biotransformation of β-Lactam Antibiotics

The β-lactam compounds have been the most amenable substrates for useful antibiotic biotransformation using microorganisms and their enzymes. Of these transformation reactions the deacylation of the penicillins and the acylation of 6-aminopenicillanic acid (6-APA), 7-aminocephalosporanic acid (7-ACA), and 7-amino-3-deacetoxycephalosporanic acid have been of prime importance. In general the penicillin acylation and deacylation reactions are catalyzed by a single type of enzyme with its directional equilibrium controlled by reaction pH.[60] The most notable exception is an alternate mode of acylation of 6-APA catalyzed by an enzyme from *Penicillium chrysogenum* which requires acyl-CoA.[57] The application of cephalosporin deacylation to the preparation of 7-ACA for synthetic purposes has been limited by narrow substrate specificities. Recently, enzyme catalyzed deacylation of nocardicin C to 3-aminonocardicinic acid has been reported.[95] Specific examples of deacylation and acylation reactions are cited in Table 1.

In addition to the 7-position acylation or deacylation of either 7-aminocephem compounds or some cephalosporins, other cephalosporin biotransformations involving acetylation or deacetylation at position 3 are known to occur (Table 1). One of these reactions, the acetylation of deacetylcephalosporin C, is biosynthetic in nature and is catalyzed by an acetyl-CoA requiring enzyme from *Cephalosporium acremonium*.[56] Examples of deacetylation reactions include the conversions of cephalosporin C to deacetylcephalosporin C by *Bacillus cereus* and cephalothin to deacetylcephalothin by *Escherichia coli*.[37,135]

All of the previously mentioned biotransformations of β-lactam antibiotics result in

Table 1
BIOTRANSFORMATION OF β-LACTAM ANTIBIOTICS

Substrate	Transformation product	Biological agent	Ref.
1. Deacylation			
Penicillin G	6-Aminopenicillanic acid (6-APA) and phenylacetic acid	*Penicillium chyrsogenum* and *Aspergillus oryzae*	165
Penicillin G	6-APA and phenylacetic acid	*Escherichia coli*	161
Penicillin G	6-APA and phenylacetic acid	*Alcaligenes faecalis*	29
10 Penicillins	6-APA and respective acyl groups	*P. chrysogenum*[c]	30
Penicillin V	6-APA and phenoxyacetic acid	*Cephalosporium*, CMI49137[c]	30
5 Penicillins	6-APA and respective acyl groups	*A. faecalis*	30
4 Penicillins	6-APA and respective acyl groups	*E. coli*[c]	30
10 Penicillins	6-APA and respective acyl groups	*Bacillus megaterium*[c]	27
Penicillin G and penicillin V	6-APA and respective acyl groups	*Fusarium semitectum*	23
20 Penicillins	6-APA and respective acyl groups	*Proteus rettgeri, A. faecalis, Erwinia aroideae, Achromobacter* sp., *Kluyvera citropila, Pseudomonas melanogenum, Micrococcus roseus, Streptomyces lavendulae*, and *Nocardia* sp.	168
Cephalosporin G	7-Aminocephalosporanic acid (7-ACA) and phenylacetic acid	*Streptomyces* sp.,[c] and *Achromobacter* sp.[c]	227
Phenoxyacetylcephalosporin and phenoxymercaptoacetyl cepholosporin	7-ACA and respective acyl groups	*Nocardia* sp.[c] and *Proteus* sp.[c]	68
5 Cephalosporins	7-ACA or 7-amino-3-deacetoxycephalosporanic acid (7-ADCA) and respective acyl groups	*K. citrophila, Kluyvera noncitrophila, Micrococcus luteus, Achromobacter* sp., *Flavobacterium* sp., *P. melanogenum, Micrococcus ureae, Nocardia globerula, Rhodopseudomonas sphaeroides, Pseudomonas aeruginosa*, and *Pseudomonas crusiviae*	
Nocardicin C	3-Aminonocardicinic acid and acyl group	*Pseudomonas schuylkilliensis*[c]	95
2. Acylation			
6-APA and phenylacetic acid	Penicillin G	*E. coli*	161
6-APA and phenylacetic acid	Penicillin G	*A. faecalis*	29
6-APA and phenoxyacetic acid	Penicillin V	*A. faecalis*	29
6-APA and 23 carboxylic acids	23 Penicillins	*E. coli*	32
6-APA and 7 derivatives of phenylglycine	Ampicillin	*K. citrophila*	130
6-APA and phenylacetic acid	Penicillin G	*K. citrophila*	130

Table 1 (continued)
BIOTRANSFORMATION OF β-LACTAM ANTIBIOTICS

Substrate	Transformation product	Biological agent	Ref.
6-APA and *p*-aminophenylacetic acid	*p*-Aminobenzylpenicillin	*K. citrophila*	130
6-APA and phenylmalonic acid methylether	α-Carboxybenzylpenicillin	*K. citrophila*	130
6-APA and aminocyclohexane carbonic acid methylester	Aminocyclohexylpenicillin	*K. citrophila*	130
6-APA and phenylacetic acid	Penicillin G	*E. coli*	83
6-APA and phenoxyacetylglutamic acid	Phenoxymethylpenicillin	*E. coli*	83
6-APA and *O*-chlorophenylacetylanaline	*O*-Chlorobenzylpenicillin	*E. coli*	83
6-APA and phenylglycine methylester	α-Aminobenzylpenicillin	*P. melanogenum*	142
6-APA and *p*-hydroxyphenylglycine ester	Amoxycillin	*E. coli*[c]	111
6-APA and phenylacetyl-CoA, *p*-methoxyphenylacetyl-CoA, phenoxyacetyl-*CoA, and octanoyl-CoA*	*4 Penicillins*	*P. chyrsogenum*[c]	57
7-ACA, 7-ADCA and organic acid esters	Cephalexin and cephaloglycin	*Xanthomonas citri, Xanthomonas oryzae, Acetobacter pasteurianus, Gluconobacter suboxydans, P. melanogenum, Mycoplana dimorpha,* and *Protaminobacter alboflavus*	202
7-ACA, 7-ADCA and 14 esters	20 Cephalopsorins	*K. citrophila*	187
7-ADCA and 1-(1H)-tetrazolylacetic acid	7-[1-(1H)-tetrazolylacetamido]-desacetoxycephalosporanic acid	*K. citrophila*	187
3. Acetylation (Position 3)			
Deacetylcephalosporin C and acetyl-CoA	Cephalosporin C	*Cephalosporium acremonium*[c]	56
4. Deacetylation (Position 3)			
Cephalosporin C	Deacetylcephalosporin C	*Bacillus cereus*[b]	37
Cephalothin	Deacetylcephalothin	*E. coli*	135
5. β-Lactam Ring Cleavage			
Penicillin	β-Lactam ring cleavage product	*Bacillus coli (E. coli)*	1
Penicillin G	Ring cleavage product	*Bacillus subtilis*[b]	156
Penicillins G, V and N and ampicillin	Ring cleavage products	*B. cereus*[c]	97
6-APA and penicillin G	Ring cleavage products	*Bacillus licheniformis*[c]	157
Penicillins G and V and phenoxyethyl penicillin	Ring cleavage products	*Staphylococcus aureus*[c]	159
Penicillins G and V, ampicillin and carbenicillin	Ring cleavage products	*E. coli*	159a
Penicillins G and V, and ampicillin	Ring cleavage products	*Streptomyces,* K11[c]	70
Penicillins G and V, and ampicillin	Ring cleavage products	*Streptomyces albus*[c]	70

Table 1 (continued)
BIOTRANSFORMATION OF β-LACTAM ANTIBIOTICS

Substrate	Transformation product	Biological agent	Ref.
Penicillins G and V, and ampicillin	Ring cleavage products	*Streptomyces*, K11[c]	70
Penicillin G, ampicillin and carbenicillin	Ring cleavage products	*P. aeruginosa*	201
Cephalosporin C, cephalothin, cephaloridine and cephaloglycin	Ring cleavage products	*B. cereus*[c]	97
Cephalosporin C and benzylcephalosporin C	Ring cleavage products	*B. licheniformis*[c]	157
Cephalosporin C	Ring cleavage product	*S. aureus*	159
Cephaloridine and Cephalothin	Ring cleavage products	*E. coli*[c]	159a
Cephaloridine, cephalothin and cephalosporin C	Ring cleavage products	*Enterobacter cloacae*[c]	62
Cephaloglycin and cephalothin	Ring cleavage products	*Streptomyces*, R39[c]	70
Cephaloglycin and cephalothin	Ring cleavage products	*Streptomyces*, K11[c]	70
Cephaloridine	Ring cleavage product	*P. aeruginosa*[b]	29

Note: The biological agents were microorganisms in fermentation with the following exceptions. The superscripts indicate catalysis by:

[a] Whole cells and cell-free extracts.
[b] Cell-free extracts.
[c] Purified enzymes.

the formation of compounds retaining the β-lactam ring structure. However, the most thoroughly investigated β-lactam biotransformation is β-lactam ring cleavage. Unfortunately, the ring cleavage products lack antibiotic activity and have only been superficially summarized in Table 1. The structures presented in Figure 1 are typical of penicillins and cephalosporins and give the usual numbering systems.

B. Biotransformation of Aminocyclitol Antibiotics

The aminocyclitols are a large family of antibiotics characterized by a cyclohexane ring substituted usually by two amino groups in a 1,3-relationship and several hydroxyl groups. Deoxystreptamine is the most characteristic nucleus. At least one amino sugar and commonly more than one is attached to the carbocyclic system through oxygen. Sugars are usually attached at either 4,5 or 4,6 with one sugar most frequently being a pentose. The structure of gentamicin C_{1a}, one of the components of commercial gentamicin, is shown as typical and to give the usual numbering system (see Figure 2).

For the most part, microbial modifications of aminocyclitols have been studied as a result of microbial inactivation of antibiotics by some sort of modification rather than in a planned attempt to improve activity by modification. The reactions usually involved have been either N or O acetylation, phosphorylation or nucleotidylation. An excellent review of these areas of aminocyclitol microbial modification is available.[212] However, development of mutasynthesis[185] was the result of a designed approach to the preparation of new aminocyclitols. This involves mutation of an aminocyclitol producing organism to give a mutant requiring the addition of the aminocyclohexane moiety to complete antibiotic synthesis. Addition of aminocyclohexanes which are not produced by the parent organism to the mutant fermentation gives new antibiotics. These transformations are listed in Table 2.

C. Biotransformation of Macrolide Antibiotics

The macrolide antibiotics are a numerous family characterized chemically by a large lactone ring. The classical macrolides have attached at C-5 an aminosugar although

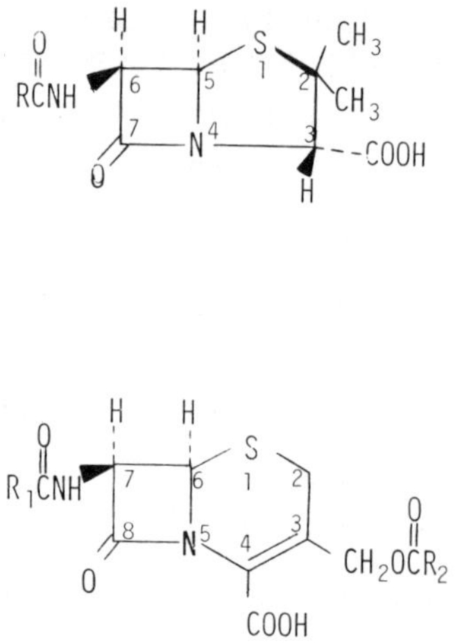

FIGURE 1. The structures of typical penicillins and cephalosporins.

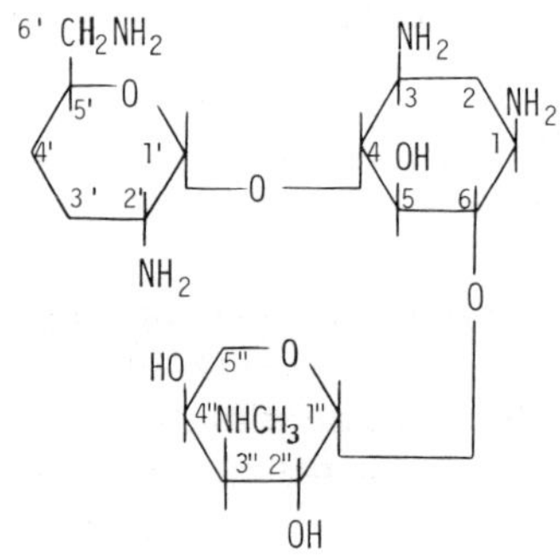

FIGURE 2. The structure of gentamicin C_{1a}.

many are known which lack this feature. Neutral sugars are commonly present attached at either C-3 or at C-4′. The structure of erythromycin A (or simply erythromycin) is shown as being typical of this family and to indicate the numbering system (see Figure 3).

Microbial modification of members of this family has not been extensive and has consisted for the most part of acylation and deacylation. Reduction of aldehyde and ketone carbonyls has been done as well as the reverse, i.e., oxidation of a secondary alcohol to a ketone. Other oxidations have involved hydroxylation at a tertiary carbon and epoxidation of an olefin. The most interesting microbial modification of this group, although not strictly using an antibiotic as a substrate, is the glycosylation of an aglycone, narbonolide, by *Streptomyces platensis* subsp. *malvinus* MCRL0388,[108] a platenomycin producer.

D. Biotransformation of Anthracycline Antibiotics

Microbial biotransformation of the polyketide derived anthracycline antibiotics has been accomplished via ketonic carbonyl reduction,[151a,230a] reductive glycosidic cleavage, oxidation, acylation, and glycosidation, as well as by the inhibition of methylation and esterification. These transformation reactions have been catalyzed using whole cells, cell-free extracts, and purified enzymes.

Table 2
BIOTRANSFORMATION OF AMINOCYCLITOL ANTIBIOTICS

Substrate	Transformation product	Biological agent	Ref.
1. Transglycosylation			
Neamine	5-Glucosylneamine	Clarase[c]	46
Kanamycin A	GlucosylkanamycinA		46, 47
Gentamicin	Glucosylgentamicin		46
Paromomycin	Glucosylparomomycin		46
Validoxylamine A	Validamycin A	*Rhototorula glutinis, Rhodotorula marina, Rhodotorula lactosa, Rhodotorula rubra*	74 74 74 74
2. Hydrolysis			
Mannosidostreptomycin	Streptomycin	*Streptomyces griseus*	154
Validamycin A	Validoxylamine A	*Pseudomonas, Erwina aroidea, Corynebacterium aquaticum, Oospora destructa, Rhizopus stolonifer, Symcephalastrum* sp.	73,74
Validamycin B	Hydroxyvalidamine	*Pseudomonas denitrificans*	73
Validamycin C	Validamycin A	*Endomycopsis fibuliger*	74
Validamycin E			
Validamycin F		*Endomycopsis chodatii, Endomyces decipiens, Candida intermedia*	74
Validamycin D	Validoxylamine	*E. fibuliger,*	74
		E. chodatii,	74
		E. decipiens,	74
		C. intermedia,	74
		P. denitrificans,	74
		Pseudomonas maltophila,	74
		Pseudomonas melanogenum,	74
		Candida sp.	74
3. *N*-Acetylation			
Amikamicin	6′-*N*-Acetylamikamicin	*Pseudomonas aeruginosa*[c] GN315, *Klebsiella pneumoniae,*[b] *Streptomyces kananyceticus,*[b] *Staphylococcus aureus*[c] R. Palm	236 121 58 103
Butirosin A	6′-*N*-Acetylbutirosin A	*P. aeruginosa* GN315[c]	236
Butirosin B	6′-*N*-Acetylbutirosin B	*S. aureus* R. Palm[c]	236
Butirosin	2′-*N*-Acetylbutirosin	*Providencia* GN1554[c]	242
3′, 4′-Dideoxykanamycin B	6′-*N*-Acetyl-3′,4′-dideoxy-kanamycin B	*P. aeruginosa* GN4925,[c]	81
		P. aeruginosa GN315[c]	82,239
Gentamicin A	6′-*N*-Acetylgentamicin A	*S. kananyceticus*[b]	16
Gentamicin B	6′-*N*-Acetyl-gentamicin B	*S. aureus* R. Palm[c]	103
Gentamicin C_1	3-*N*-Acetyl-gentamicin C_1	*P. aeruginosa* PST,	19
		P. aeruginosa 130	20
	6′-*N*-Acetyl-gentamicin C_1	*Pseudomonas*	58
		S. kananyceticus,[b]	58
		P. aeruginosa GN315,[c]	236
		Escherichia coli[c] K12C66OR135	220

Table 2 (continued)
BIOTRANSFORMATION OF AMINOCYCLITOL ANTIBIOTICS

Substrate	Transformation product	Biological agent	Ref.
Gentamicin C_{1a}	6′-*N*-Acetyl-gentamicin C_{1a}	*E. coli* W677,[b]	15
		P. aeurginosa 3796,[b]	58
		Moraxella 731 and 2513,[c]	101
		S. aureus R. Palm,[c]	103
		K. pneumoniae 355,[b]	121
		P. aeruginos GN315[c]	236
	2′-*N*-Acetylgentamicin C_{1a}	*Streptomyces spectabilis,*[b]	16
		S. kanamyceticus,[b]	16
		Providencia 164,[c]	26
		Providencia GN1554[c]	242
	3-*N*-Acetylgentamicin C_{1a}	*E. coli,*[b]	104
		E. coli K12C 600R135,[c]	220
		P. aeruginosa 130[b]	24
	Acetylgentamicin C_{1a}	*P. aeruginosa* 99 and Cope 18[b]	122
Gentamicin C_2	6′-*N*-Acetylgentamicin C_2	*K. pneumoniae* 355,[b]	121
		P. aeruginosa GN315,[c]	236
		P. aeruginosa 99 and Cope 18[b]	122
Gentamicin	3-*N*-Acetylgentamicin	*E. coli* C-600JR88[c]	234
6′-*N*-Acetylgentamicin C_{1a}	3,6′-Di-*N*-Acetylgentamicin C_{1a}	*P aeruginosa* 130[b]	24
Kanamycin A	3-*N*-Acetylkanamycin A	*P. aeruginosa* PST,[c]	19
	6′-*N*-Acetylkanamycin A	*E. coli W677,*[b]	15
		S. Kanamyceticus,[b]	16
		P. aeruginosa 33796,[b]	58
		P. aeruginosa GN4925,[c]	81
		P. aerubinosa GN315,[c]	82, 244
		Moraxella 731 and 2513,[c]	101
		S. aureus R. Palm,[c]	103
		K. pneumoniae 355,[b]	121
		E. coli K12[b]	215
Kanamycin B	3-*N*-Acetylkanamycin B	*P. aeruginosa*[c]	19
		PST,	15
	6′-*N*-Acetylkanamycin B	*E. coli* W677,[b]	16
		S. kanamyceticus,[b]	58
		P. aeruginosa 3796,[b]	81
		P. aeruginosa GN4925,[c]	
		Moraxella 713 and 2513,[c]	103
		S. aureus R. Palm,[c]	103
		K. pneumoniae 355,	241
		P. aeruginosa GN315,	236
		Providencia GN1554[c]	242
Kanamycin C	6′-*N*-Acetylkanamycin C	*P. aeruginosa* 3796,[b]	58
		P. aeruginosa GN315,[c]	236
		Providencia GN1554,[c]	242
		S. kanamyceticus[b]	16
Lividomycin A	2′-*N*-Acetyllividomycin A	*Providencia*[c] GN1554	242

Table 2 (continued)
BIOTRANSFORMATION OF AMINOCYCLITOL ANTIBIOTICS

Substrate	Transformation product	Biological agent	Ref.
Kanamycin C	6′-*N*-Acetylkanamycin C	*P. aeruginosa* 3796,[b]	58
		P. aeruginosa GN315,[c]	236
		Providencia GN1554,[c]	242
		S. kanamyceticus[b]	16
Lividomycin A	2′-*N*-Acetyllividomycin A	*Providencia*[c] GN1554	242
6′-*N*-Methylkanamycin B	6′-*N*-Acetyl-6′-*N*-methylkanamycin B	*P. aeruginosa* GN4925[c]	81
Neamine	6′-*N*-Acetylneamine	*Moraxella* 2513 and 731[c]	101
Neomycin	6′-*N*-Acetylneomycin	*Moraxella* 2513 and 731[c]	101
Neomycin B	3-*N*-Acetylneomycin B	*P. aeruginosa*	19
		Streptomyces fradiae	160
	6′-*N*-Acetylneomycin B	E. coli W677,[b]	15
		S. kanamyceticus,	16
		P. aeruginosa 3796,[b]	58
		S. aureus R. Palm[c]	103
Neomycin C	3-*N*-Acetylneomycin C	*S. fradiae*	160
	6′-*N*-Acetylneomycin C	*S. aureus* R. Palm[c]	103
Paromamine	6′-*N*-Acetylparmoamine	*P. aeruginosa* GN315[c]	236
Paromomycin	3-*N*-Acetylparomomycin	*E. coli* R176,[b]	104
		P. aeruginosa PST[c]	19
	6′-*N*-Acetylparomomycin	*S. kanamyceticus*	16
Ribostamycin	3-*N*-Acetylribostamycin	*Streptomyces ribosidificus,*	92
	2′-*N*-Acetylribostamycin	*Providencia* GN1554,[c]	242
	6′-*N*-Acetylribostamycin	*S. aureus* R. Palm,[c]	103
		P. aeruginosa[c] GN315	236,244
Sisomicin	3-*N*-Acetylsisomicin	*P. aeruginosa* PST,[b]	19
	2′-*N*-Acetylsisomicin	*P. aeruginosa* 130[b]	24
	6′-*N*-Acetyl-sisomicin	*Providencia* 164,[c]	26
		S. kanamyceticus,	16
		P. aeruginosa 3796,[b]	58
		Moraxella 2513,[c]	101
		K. pneumoniae 355,[b]	121
		P. aeruginosa GN269,[b]	140
		P. aeruginosa NG150[b]	141
Tobramycin	3-*N*-Acetyltobramycin	*P. aeruginosa* PST,[c]	19
		E. coli R176,[b]	104
	2′-*N*-Acetyltobramycin	*Providencia* 164,[c]	26
	6′-*N*-Acetyltobramycin	*S. kanamyceticus,*	16
		P. aeruginosa 3796,[b]	58
		Moraxella 2513,[c]	101
		K. pneumonia 355,[b]	121
		P. aeruginosa GN315	236
4. Phosphorylation			
Amikamicin	3′-*O*-Amikamicin phosphate	*S. aureus* RN450(S),	34
	2″-Amikamicin phosphate	209P and RN450[b]	
		S. aureus R. Palm[c]	103

Table 2 (continued)
BIOTRANSFORMATION OF AMINOCYCLITOL ANTIBIOTICS

Substrate	Transformation product	Biological agent	Ref.
4. Phosphorylation			
Amikamicin	3′-*O*-Amikamicin phosphate	*S. aureus* RN450(S), 209P and RN450[b]	34
	2″-Amikamicin phosphate	*S. aureus* R. Palm[c]	103
Butirosin	3-*O*-Butirosin phosphate	*E. coli* JR66,[c]	25
		P. aeruginosa Ps49,[c]	25
		P. aeruginosa 21-75[c]	221
Butirosin A	3′-*O*-Butirosin A phosphate	*E. coli* JR61/W677,[c]	240
		P. aeruginosa[c] TI-13 and B13,	117
		Bacillus[c] *circulans*	116
Dihydrostreptomycin	3″-*O*-Dihydrostreptomycin phosphate	*P. aeruginosa*[c] TI-13,	80,91, 123
		S. griseus[c] ATCC10971,	236
	3′-(2-Hydroxymethyl-*O*-dihydrostreptomycin phosphate	*Streptomyces bikiniensis*[b] ATCC11062,	119, 226
		Streptomyces[b]	225
3″-*O*-Dihydrostreptomycin phosphate	3″,6-*O*-Dihydrostreptomycin diphosphate	*S. bikiniensis* ATCC11062[b]	226
Gentamicin A	2″-*O*-Gentamicin A phosphate	*S. aureus* R. Palm[c]	103
Gentamicin B	2″-*O*-Gentamicin B phosphate		103
Gentamicin C_1	2″-*O*-Gentamicin C_1 phosphate		103
Gentamicin C_{1a}	2″-*O*-Gentamicin C_{1a} phosphate		103
Gentamicin C_2	2″-*O*-Gentamicin C_2 phosphate		103
3′-(2-Hydroxymethyl-*O*-dihydrostreptomycin phosphate	6-*O*-3′-(2-Hydroxymethyl-*O*-streptomycin diphosphate	*S. bikiniensis* ATCC11062[b]	226
Kanamycin A	3′-*O*-Kanamycin A phosphate	*P. aeruginosa*[c] 21-75,	221
		P. aeruginosa[c] H9,	40, 117, 213
		P. aeruginosa[c] TI-13 and B-13,	117
		E. coli,[c]	96
		E. coli K12ML1629,[c]	123, 214
		E. coli Jr61/W677,[c]	240
		S. aureus B294 and B295[c]	41
	2″-*O*-Kanamycin A phosphate	*S. aureus* R. Palm[c]	103
Kanamycin B	3′-*O*-Kanamycin B phosphate	*P. aeruginosa*,[c]	39
		P. aeruginosa[c] H9,	40
		P. aeruginosa GN573,	147
		P. aeruginosa 21-75	221
	2″-*O*-Kanamycin B phosphate	*S. aureus* R. Palm[c]	103
Kanamycin C	3′-*O*-Kanamycin C phosphate	*P. aeruginosa* H9[c]	40

Table 2 (continued)
BIOTRANSFORMATION OF AMINOCYCLITOL ANTIBIOTICS

Substrate	Transformation product	Biological agent	Ref.
Neamine	3′-*O*-Neamine phosphate	*E. coli*[c] JR611/W677	240
Nebramine	3′-*O*-Nebramine phosphate	*P. aeruginosa* GN573	147
Neomycin	3′-*O*-Neomycin phosphate	*P. aeruginosa* H9,[c]	40
		P. aeruginosa[c]	221
Neomycin B	3′-*O*-Neomycin B phosphate	*E. coli* JR39,[c]	25
		E. coli JR66,[c]	25
		P. aeruginosa Ps49,[c]	25
		P. aeruginosa GN573	147
Paromamine	3′-*O*-Paromamine phosphate	*E. coli* K12ML1629,[c]	96
		P. aeruginosa NIH JB-328[c]	106, 123, 214
Paromomycin	3′-*O*-Paromomycin phosphate	*P. aeruginosa* GN573	147
Ribostamycin	3′-*O*-Ribostamycin phosphate	*E. coli* JR61/W677[c]	212, 240
	5″-*O*-Ribostamycin phosphate	*P. aeruginosa*[b] GN573	85
Sisomicin	2″-*O*-Sisomicin phosphate	*S. aureus*[c] R. Palm	103
Streptomycin	6-*O*-Streptomycin phosphate	*S. griseus* ATCC12475 and SC2376,	120
		S. griseus HUT6037	132, 133, 137, 138
	3′-*O*-Streptomycin phosphate	*E. coli* with R factors,	151
	Streptomycin phosphate	*P. aeruginosa*[c] H1-H9 and A3,	40
		P. aeruginosa TI-13,[b]	90
		S. bikiniensis ATCC11062[b]	119
	3″-*O*-Streptomycin phosphate	*S. griseus* ATCC10971	226
3″-*O*-Streptomycin phosphate	6,3″-*O*-Streptomycin diphosphate	*S. bikiniensis* ATCC11062[b]	119
Tobramycin	2″-*O*-Tobramycin phosphate	*S. aureus*[c] R. Palm	103
5. Nucleotidylation			
Amikamicin	4′-*O*-Adenylylamikamicin	*S. aureus* ApO1,[b]	102
		S. aureus RAPO1	103
Butirosin A	4′-*O*-Adenylylbutirosin A	*S. aureus* ApO1,[b]	102
		S. aureus RAPO1	103
Butirosin B	4′-*O*-Adenylylbutirosin B	*S. aureus* ApO1,[b]	102
		S. aureus RAPO1	103
3′,4′-Dideoxykanamycin B	2″-*O*-Adenylyl-3′,4′-dideoxykanamycin B	*E. coli* JR66/W677[c]	124,237
	2″,*O*-guanosylyl-3′,4′-dideoxykanamycin B	*E. coli* JR66/W677[c]	238
	2″-*O*-Inosylyl-3′,4′-dideoxykanamycin B	*E. coli* JR66/W677[c]	238

Table 2 (continued)
BIOTRANSFORMATION OF AMINOCYCLITOL ANTIBIOTICS

Substrate	Transformation product	Biological agent	Ref.
Gentamicin A	4′-*O*-Adenylylgentamicin A	*S. aureus* ApO1,[b]	102
		S. aureus RAPO1	103
	O-Adenylylgentamicin A	*E. coli* JR66/W677[c]	18
Gentamicin B	4′-*O*-Adenylylgentamicin B	*S. aureus* ApO1,[b]	102
		S. aureus RAPO1	103
Gentamicin C_1	*O*-Adenylylgentamicin C_1	*P. aeruginosa* POW,[b] *E. coli* JR66/W677,[c] *E. coli* W677/HJR66[c]	72
Gentamicin C_{1a}	*O*-Adenylylgentamicin C_{1a}	*P. aeruginosa* POW,[b] *E. coli* JR66/W677,[c] *E. coli* JR66/HJR66[c]	18
Gentamicin C_2	*O*-Adenylylgentamicin C_2	*P. aeruginosa* POW,[b] *E. coli* JR66/W677,[c] *E. coli* JR66/HJR66[c]	193
Kanamycin A	*O*-Adenylylkanamycin A	*E. coli* JR66/W677,[c]	18
		P. aeruginosa POW,[b]	72
		E. coli K12ML1629,[c]	143
		E. coli W677/HJR66[b]	193
	4′-*O*-Adenylylkanamycin A	S. aureus ApO1,[b]	102
		S. aureus RAPO1	103
Kanamycin B	*O*-Adenylylkanamycin B	*P. aeruginosa* POW,[b]	72
		E. coli W677/HJR66[b]	193
	4′-*O*-Adenylylkanamycin B	*S. aureus* ApO1,[b]	102
		S. aureus RAPO1	103
Lividomycin	4′-*O*-Adenylyllividomycin	*S. aureus* ApO1,[b]	102
		S. aureus RAPO1	103
Neomycin B	4′-*O*-Adenylylneomycin B	*S. aureus* ApO1,[b]	102
		S. aureus RAPO1	103
Neomycin C	4′-*O*-Adenylylneomycin C	*S. aureus* ApO1,[b]	102
		S. aureus RAPO1	103
Paromamine	*O*-Adenylylparomamine	*E. coli* K12ML1629[c]	143
Paromomycin	4′-*O*-Adenylylparomomycin	*S. aureus* ApO1,[b]	102
		S. aureus RAPO1	103
Ribostamycin	4′-*O*-Adenylylribostamycin	*S. aureus* ApO1,[b]	102
		S. aureus RAPO1	103
Sisomicin	*O*-Adenylylsisomicin	*P. aeruginoosa* POW[b]	72
Spectinomycin	9-*O*-Adenylylspectinomycin	*E. coli*[c] NR73/W677,	17
	O-Adenylylspectinomycin	*S. aureus*[b] MS7990 and MS9263	79
Streptomycin	3″-*O*-Adenylylstreptomycin	*E. coli* NR73/W677,[c]	17
		E. coli K12ML1629,[c]	143, 203, 219
		E. coli R factor-resistant[c]	241
	O-Adenylylstreptomycin	*S. aureus* MS27 and MS7990[b]	79

Table 2 (continued)
BIOTRANSFORMATION OF AMINOCYCLITOL ANTIBIOTICS

Substrate	Transformation product	Biological agent	Ref.
Tobramycin	4′-*O*-Adenylyltobramycin	*P. aeruginosa* POW,[b]	72
		E. coli W677/HJR66[b]	193
6. Methylation			
Gentamicin C_{1a}	Gentamicin C_{2b}	*Micromonospoora purpurea*-Paro 346	206
Gentamicin C_2	Gentamicin C_1		206
7. Mutasynthesis[e]			
Streptidine	New product	*B. circulans* mutant	28
Streptamine	2-Hydroxybutirosin		28, 204
2,5-Dideoxystreptamine	5-Deoxybutirosamine		204
3′-Chloro-3′-deoxyneamine	3′-Chloro-3-deoxybutirosin A	*Bacillus botulinus* GD-72	64
Streptamine	2-Hydroxygentamicins C_1, C_{1a}, and C_2	*M. purpura* NRRL2953	162
2,5-Dideoxystreptamine	5-Deoxygentamicins C_1, C_{1a}, and C_2		162
3-Epistreptamine	New products		162
2,4,6/3,5-Pentahydroxycyl ohexanone	2-Hydroxygentamicins C_1, C_{1a}, and C_2	*M. purpurea* VIf-3P	35, 36
Streptamine	Hybrimycins A_1 and A_2	*S. fradiae* 3535	185, 186
2-Epistreptamine	Hybrimycins B_1 and B_2		185, 186
Streptamine	Mutamicin 1	*Micromonospora inyoensis* NRRL3292	207
2,5-Dideoxystreptamine	Mutamicins 2 and 2a		207
2-Deoxystreptamine	Sisomicin	*M. inyoensis* 1550F	205
Neamine			205
Paromamine			205
Neamine	Ribostamycin	*Streptomyces ribosidificus* AF-1	94
3′,4′-Dideoxyneamine	3′,4′-Dideoxyribostamycin		94
Streptamine	2-Hydroxyribostamycin		94
2-Epistreptamine	2-Epihydroxyribostamycin		94
1-*N*-Methyldeoxystreptamine	1-*N*-Methylribostamycin		94
	6′-Deamino-6′-hydroxyl-1-*N*-methylkanamycin A	*S. kanamyceticus* mutant	94
2-Epistreptamine	6′-Deamino-6′-hydroxyl-2-epihydroxykanamycin A		94
2-Deoxystreptidine	Streptomutin A	*S. griseus* MITA5	94
8. Multiple conversions			
Gentamicin A	Gentamicins C_1, C_{1a}, and C_2	*M. purpura* Paro 346	206
Gentamicin A_1	Antibiotic 66-40D	*M. inyoeneis* 1550F	205
Gentamicin A_2	Sisomicin		205
Gentamicin X_2	Gentamicins C_1, C_{1a}, and C_2	*M. purpura* Paro 346,	206
	Sisomicin	*M. inyoeneis* 1550F	205
Antibiotic JI-20A	Gentamicins C_{1a} and C_{2b}	*M. purpura* Paro 346,	206
	Sisomicin	*M. inyoeneis* 1550F	205
Antibiotic JI-20B	Gentamicins C_1 and C_2	*M. purpura* Paro 346	206
Antibiotic G-418	Gentamicins C_1 and C_2		206
Antibiotic 66-40B	Sisomicin	*M. inyoeneis* 1550F	205

Note: The biological agents were microorganisms in fermentation with the following exceptions. The superscripts indicate catalysis by:

[a] Whole cells and cell extracts.
[b] Cell extracts.
[c] Purified enzyme.
[d] Whole cells and purified enzyme.
[e] All of the organisms reported in this section were mutants lacking the ability to synthesize the normally produced diaminocyclohexitol.

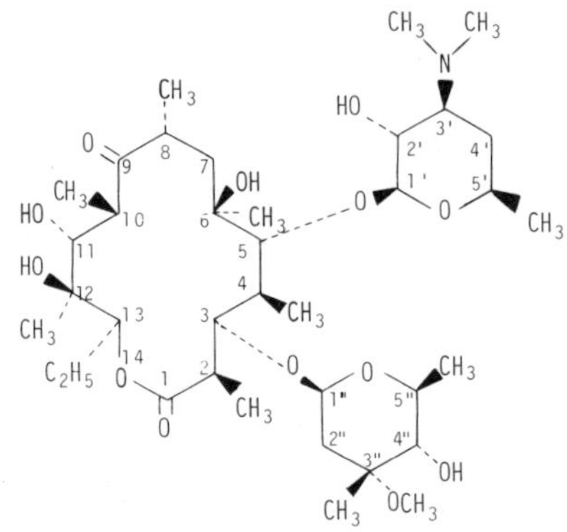

FIGURE 3. The structure of erythromycin A.

Ketonic carbonyl reductions have involved the C-13 carbonyl of daunomycin-adriamycin type anthracyclines, as well as the C-10 ketonic carbonyl of steffimycin related compounds. Specific examples of these conversions are cited in Table 4.

In the instance where microbial C-13 carbonyl reduction of daunomycin was investigated biochemically, NADPH was required as a co-factor as was also the case concerning the identical reaction catalyzed by mammalian microsomal preparations.[112,49] Microbial conversion of 7-deoxydaunomycinone and 7-deoxyadriamycinone to 7-deoxydaunomycinol aglycone and 7-deoxyadriamycinol aglycone was also shown to require NADPH.[114,112]

Reductive cleavage of anthracyclines or anthracyclinones results in the formation of 7-deoxyanthracyclinones and free sugar moieties in the case of anthracycline substrates or 7-deoxyanthracyclinones, and H_2O in the case of anthracyclinone substrates. Examples of such conversions are presented in Table 4. When investigated biochemically, these microbial conversions were shown to require NADH,[112-114,164,231] whereas the mammalian counterpart of this reaction requires NADPH.[12] Both the mammalian and bacterial enzymes catalyzing this reaction have been purified,[118,144] and in the case of the mammalian system, the enzyme was identified as a NADPH linked flavoprotein reductase.

A specific oxidoreductase which catalyzes the conversion of aclacinomycin A to aclacinomycin Y was purified from culture filtrates of *Streptomyces galilaeus*.[246] The enzyme catalyzes the oxidation of the terminal sugar moiety of aclacinomycin A by the removal of 2 electrons. This example of anthracycline oxidative bioconversion is presented in Table 4.

In addition to the biotransformation mechanisms involving oxidation and reduction, anthracycline biotransformation has been performed by means of acylation and glycosidation, as well as by the inhibition of methylation and esterification. Specific examples of these conversions are presented in Table 4. In these studies daunomycin and daunomycinol were converted to their N-acetyl derivatives using *Bacillus cereus* var. *mycoides*[59] and various anthracyclinones were glycosylated by *S. galilaeus*,[223] *Streptomyces coeruleorubidus*,[20] and *Streptomyces aureofaciens*.[115] In addition, DL-ethionine was shown to inhibit methylation and esterification reactions performed by *S. coeruleorubidus*.[19a]

The structure of aclacinomycin A is presented as typical of anthracycline antibiotics (Figure 4). This structure is numbered in the conventional manner for members of the anthracycline class of antibiotics.

Table 3
BIOTRANSFORMATION OF MACROLIDE ANTIBIOTICS

Substrate	Transformation product	Biological agent	Ref.
1. Acylation			
4″-*O*-Depropionylmaridomycon III	Maridomycin V	*S-reptomyces* . Strain no. K-342	222
Lankacidin C + Acetate	Lankacidin A	*Aspergillus sojae*[c]	54
Lankacidin C + formate	14-*O*-Formyllankacidin A	*Aspergillus niger*[c]	54
Lankacidin C + propionate	14-*O*-Propronyl lankacidin A	*Trametes sanguinea*[c]	54
Lankacidinol	Lankacidinol A		54
Lankacidinol + acetate	Lankacidin A	*Streptomyces*[b]	61
	2′,14-Di-*O*-Acetyllankacidin	*rochei* var. *volubilis*	61
	Lankacidinol A		61
Lankacidinol + propionate	14-*O*-Propionyllackacidinol		61
Lankacidin C + butyrate	8-*O*-Butyryllankacidin C	*Bacillus megaterium* IF012108[d]	126
Lankacidin C + Isobutyrate	8-*O*-Isobutyryllankacidin C		126
Lankacidin C + valerate	8-*O*-Valeryllankacidin C		126
Lankacidin C + isovalerate	8-*O*-Isovaleryllankacidin C		126
Leucomycin A_1	Leucomycin A_3	*Streptomyces kitasatoensis* Strain No. 66-14-3	88, 148, 149
Leucomycin U			88
Leucomycin V	Leucomycin U	*S. kitasatoensis* Strain No. 66-14-3,	88
		Streptomyces ambofaciens KA-1028	88
	Leucomycin A_3	*S. kitasatoensis* Strain No. 66-14-3	88
2. Deacylation			
8-*O*-Benzoyl-14-*O*-acetyllankacidin C	8-*O*-Benzoyllankacidin C	*S. rochei* var. volubilis[b]	61
8-*O*-Butyryl-14-*O*-acetyllankacidin C	8-*O*-Butyryllankacidin C		61
Carbomycin A	4″-*O*-Deisovalerylcarbomycin A	*Cunninghamella, Helicostylum, Mortierella,*	208
Carbomycin B	4″-*O*-Deisovalerylcarbomycin B	Monilia, Pencillium, Sporotrichum, Trichoderma, Pestalotiopsis	208
8-*O*-Crotonyl-14-*O*-acetyllankacidin C	8-*O*-Crotonyllankacidin C	*S. rochei* var. *volubilis*[b]	61
4″-*O*-Depropionyl-10-dihydromaridomycin III	A_4	*Streptomyces* sp. strain No. 11-245	184
4″-*O*-Depropionylmaridomycin III	A_3		184
10-Dihydromaridomycin III	A_2 and A_4		184
8,14-Di-*O*-acetyllankacidin C	8-*O*-Acetyllankacidin C	*A. sojae*,[c] *A. niger*,[c] *T. sanguinea*,[c]	54
		S. rochei var. *volubilis*[b]	61
8,14-Di-*O*-butyryllankacidin C	8-*O*-Butyryllankacidin C	*B. megaterium*[d] IF012108	126
8,14-Di-*O*-propionyllankacidin C	8-*O*-Propionyl lankacidin C	*A. sojae*,[c] *A. niger*,[c] *T. sanguinea*,	54
		S. rochei var. *volubilis*[b]	61

Table 3 (continued)
BIOTRANSFORMATION OF MACROLIDE ANTIBIOTICS

Substrate	Transformation product	Biological agent	Ref.
Lankacidin A	Lankacidin C	*A. sojae,*[c] *A. niger,*[c] *T. sanguinea*[c]	54
		S. rochei var. *volubilis*[c]	63
Lankacidinol A	Lankacidinol	*A. niger,*[c]	54
		A. sojae,[c]	53, 54
		T. sanguinea,[c]	53, 54
		S. rochei var. *volubilis*[c]	53
Leucomycin A_1	4″-*O*-Deisovalerylleucomycin A_1	*Cunninghamella, Helicostylum, Mortierella, Monilia Pencillium*	208
Leucomycin A_1, and A_3 - A_7	Products a and b (deacylated)	*Mucor spinescens* IAM6071	191
Leucomycin A_3	4″-*O*-Deisovalerylleucomycin A_3	*B. megaterium* 91277,[b]	127
		Cunninghamella, Helicostylum, Mortierella, Monilia, Penicillium, Sporotrichum, Trichoderma, Pestalotiopsis	208
Maridomycin III	4″-*O*-Depropionylmaridomycin III	*B. megaterium* 91277,[b]	127
		Streptomyces pristinaespiralis IF013074, *Streptomyces olivaceus* 219,	129
		Streptomyces lavendulae strain No. K-122, *Streptomyces* sp. strain No. K-245	183
			184
	4″-*O*-Depropionyl-18-dihydromaridomycin III	*S. lavendulae* strain No. K-122,	183
		Streptomyces sp. strain No. K-245	184
	A_1 and A_3		184
8-*O*-Nicotionoyl 14-*O*-acetyllankacidin C	8-*O*-Nicotinoyllankacidin C	*S. rochei* var. *volubilis*[b]	61
Niddamycin	4″-*O*-Deisovalerylniddamycin	*Cunninghamella, Helicostylum, Montierella, Monilia, Penicillium, Sporotrichum, Trichoderma, Pestalotiopsis*	208
8-*O*-Propionyl-14-*O*-acetyllankacidin C	8-*O*-Propionyllankacidin C	*S. rochei* var. *volubilis*[b]	61
9-*O*-Propionylmaridomycin III	4″-O-Depropionylmaridomycin III	*S. pristinaespiralis* IF013074,	129
	4″-*O*-Depropionyl-9-*O*-propionylmaridomycin III	*B. megaterium,*[b] *S. olivaceus* 219	127
			129
2′,4,14-Tri-*O*-acetyllankacidional	8-*O*-Acetyllankacidin C	*S. rochei*[b] var. *volubilis*	61
	8-*O*-Acetyllankacidinol		
	2′,8-Di-*O*-acetyllankacidinol		61
			61
SF 837 A_2 Acylated derivatives	Deacylated SF 837 A_2 Derivatives	Organisms of eleven genera	131, 191

Table 3 (continued)
BIOTRANSFORMATION OF MACROLIDE ANTIBIOTICS

Substrate	Transformation product	Biological agent	Ref.
8-*O*-Valeryl-14-*O*-acetyllankacidin	8-*O*-Valeryllankacidin C	*S. rochei* var. *volubilis*[b]	61
3. Oxidation			
Carbomycin B	Carbomycin A	*Streptomyces hydroscopicus* B-5050-HAM-1382	200
Lankacidinol A	Lankacidin A	Enzyme from one of *A. sojae, A. niger* or *T. sanguinea*	54
Lankacidinol		*S. rochei* var. *volubilis*[b]	61
Leucomycin A_3	3″-Hydroxyleucomycin A_3	*S. olivaceus* 219	128
	Maridomycin II	*S. hygroscopicus* B-5050-HA	200
Maridomycin	3″-Hydroxymaridomycin	*S. olivaceus* 219	128
Maridomycin II	Carbomycin A	*A. hygroscopicus* B-5050-HA	200
Narbomycin	Picromycin	*Streptomyces zarmyceticus* MCRL-0405	107
Narbomycin	Picromycin	*Streptomyces narboensis* ISP-5016	107
Narbonolide	Picromycin	*S. rochei* var. *volubilis*[b]	61
2,8,14-Tri-*O*-acetyllankacidinol	8-*O*-Acetyllankacidin	*S. rochei* var. *volubilis*[b]	61
4. Reduction			
Carbomycin A	Maridomycin II	*S. hygroscopicus* B-5050-HA	200
Carbomycin B	Leucomycin A_3		200
Maridomycin III	18-Dihydromaridomycin III	*Streptomyces* sp. strain No. K-245	184
Narbonolide	5-*O*-Mycaminosyl-9-dihydronarbonolide	*Streptomyces platensis* subsp. *malvinus* MCRL-0388 and mutant U-21	108
Tylosin	Relomycin	*S. hygroscopicus, Streptomyces griseospiralis, Nocardia cu*[illegible]*llina*	48
5. Methylation			
Erythromycin C	Erythromycin A	*Streptomyces erythreus*[c]	33
Erythromycin D	Erythromycin B	*Staphylococcus aureus* CA340	109
6. Glycosylation			
Erythronolide A Oxime	3-*O*-Oleandrosyl-5-*O*-desosaminylerythronlide A Oxime	*Streptomyces antibioticus* ATCC11891	105
Narbonolide	5-*O*-Mycaminosylnarbonolide	*S. platensis* subsp. *malvinus* MCRL-0388 and mutant U-21	108
	Narbomycin Picromycin	*S. narboensis* ISP-5016	107
7. Other Transformations			
Erythromycin A	Erythronolide, Erythralosamine, and other products	*Pseudomonas* 56[d]	50

Note: The biological agents were microorganisms in fermentation with the following exceptions. The superscripts indicate catalysis by:

[a] Whole cells and cell extracts.
[b] Cell extracts.
[c] Purified enzyme.
[d] Whole cells and purified enzyme.

E. Biotransformation of Lincomycin Related Antibiotics

The lincomycin family of antibiotics as presented in this discussion includes lincomycin, clindamycin, celesticetin, 4′-depropyl-4′-ethyl-lincomycin, S-demethyl-S-ethyllincomycin and 1′-demethyllincomycin. Microbial biotransformation of these antibiotics has been achieved through sulfoxidation, phosphorylation, nucleotidylation, demethylation, and hydroxymethylation. These conversions are summarized in Table 5. All of the biotransformation reactions cited in this section with the exception of the

Table 4
BIOTRANSFORMATION OF ANTHRACYCLINE ANTIBIOTICS

Substrate	Transformation product	Biological agent	Ref.
1. Ketonic Carbonyl Reduction			
Daunomycin	Daunomycinol (13-dihydrodaunomycin)	*Corynebacterium simplex*	134
Daunomycin	Daunomycinol	*Corynebacterium equi*	13
Daunomycin	Daunomycinol	*Mucor spinosus*[a]	112
Daunomycinone	13-Dihydrodaunomycinone (daunomycinol aglycone)	*Streptomyces aureofaciens*	75
Daunomycinone	13-Dihydrodaunomycinone	*Streptomyces coeruleorbuidus* and *Streptomyces galilaeus*	20
7-Deoxydaunomycinone	7-Deoxydaunomycinol aglycone	*Streptomyces steffisburgensis*[b]	114
7-Deoxyadriamycinone	7-Deoxyadriamycinol aglycone	*S. steffisburgensis*[b]	112
Steffimycin	10-Dihydrosteffimycin	*Actinoplanes utahensis*	233
Steffimycin B	10-Dihydrosteffimycin B	*A. utahensis* and *Chaetomium* UC®4634[a]	233
Steffimycinone	Steffimycinol (10-dihydrosteffimycinone)	*Streptomyces peucetius* var. *caesius* and *Streptomyces nogalater*[a]	232
2. Reductive Cleavage			
Adriamycin	7-Deoxyadriamycinone	*S. steffisburgensis*[b]	112
Aklavinone	7-Deoxyalkavinone	*S. coeruleorubidus*	20
Cinerubin A	Zeta-pyrromycinone	*Aeromonas hydrophila*	113
Daunomycin	7-Deoxydaunomycinone	*A. hydrophila*	113, 231
Daunomycin	7-Deoxydaunomycinone	*S. steffisburgensis*[b]	114
Daunomycinol	7-Deoxydaunomycinol aglycone	*S. steffisburgensis*[b]	112
Daunomycinol aglycone	7-Deoxydaunomycinol aglycone	*S. coeruleorubidus* and *S. galilaeus*	20
Nogalamycin	7-Deoxynogalarol	*A. hydrophila*	113, 231
Nogalamycin	7-Deoxynogalarol	*S. nogalater*[b]	164
Steffimycin	7-Deoxysteffimycinone	*A. hydrophila,*[a] *Escherichia coli,* and *Citrobacter freundii*	113, 231
Steffimycin	7-Deoxysteffimycinone	*A. hydrophila*[c]	118
Steffimycin B	7-Deoxysteffimycinone	*A. hydrophila*	113, 231
Steffimycinone	7-Deoxysteffinmycinone	*A. hydrophila*[c]	118
3. Oxidation			
Aclacinomycin A	Aclacinomycin Y	*S. galilaeus*[c]	246
4. Acetylation			
Daunomycin	N-Acetyldaunomycin	*Bacillus cereus* var. *mycoides*	59
Daunomycinol	N-Acetyldaunomycinol	*B. cereus* var. *mycoides*	59
5. Glycosidation			
Alizarin (1,2-dihydroxy)-9,10-anthraquinone	1-Hydroxy-2-(β-D-glucopyranosyloxy)-9,10-anthraquinone	*S. aureofaciens*[a]	115
Daunomycinone	Glycosides of daunomycinone	*S. coeruleorubidus*	20
Daunomycinol aglycone	Glycosides of 13-dehydrodaunomycinol	*S. coeruleorubidus*	20
ε-Isorhodomycinone	Glycosides of ε-isorhodomycinone	*S. coeruleorubidus*	20
ε-Pyrromycinone	Glycosides of ε-pyrromycinone	*S. coeruleorubidus*	20
ε-Pyrromycinone	Galirubin A	*S. galilaeus*	223
ε-Rhodomycinone	Glycosides of ε-rhodomycinone	*S. coeruleorubidus*	20

Table 4 (continued)
BIOTRANSFORMATION OF ANTHRACYCLINE ANTIBIOTICS

Substrate	Transformation product	Biological agent	Ref.
6. Inhibition of Methylation			
DL-Ethionine (inhibitor of anthracyclinone 4-O-methylation)	Carminomycinones	*S. coeruleorubidus*	19a
7. Inhibition of Esterification			
DL-Ethionine (inhibitor of anthracyclinone 10-COOH-esterification)	Carminomycinones	*S. coeruleorubidus*	19a

Note: The biological agents were microorganisms in fermentation with the following exceptions. The superscripts indicate catalysis by:

[a] Whole cells and cell-free extracts.
[b] Cell-free extracts.
[c] Purified enzymes.

phosphorylation of clindamycin were performed using whole cells in fermentation. The exception was catalyzed additionally by a cell-free extract of *Streptomyces coelicolor*.[31] The structure of lincomycin is shown as typical and to give the usual numbering system (Figure 5).

F. Biotransformation of Nucleoside Antibiotics

The nucleoside antibiotics are a rather large family which can be divided into two large classes, the N-nucleosides exemplified by psicofuranine (1) and the C-nucleosides exemplified by formycin (2). They are made up of a sugar glycosidically bonded to a heterocyclic ring with, in many cases, other groups attached at various positions. A number of different microbial conversions have been carried out on nucleoside antibiotics but only on a very limited number of compounds.

G. Biotransformation of Actinomycins and Peptide Antibiotics

Biomodification of the actinomycins has been achieved primarily through precursor directed or controlled biosynthesis. The precursors which are related to the amino acid moieties of the actinomycins are supplied to the actinomycin producing cultures during fermentation. Specifically, the precursors employed have been branched chain amino acids, norvaline, proline, proline analogs, pipecolic acid, sarcosine, glycine, as well as azetidine-2-carboxylic acid. As a result of these precursor additions there has been enhanced formation of actinomycins normally produced in trace amounts or synthesis of novel actinomycins. The results of these studies are summarized in Table 7.

Bioconversion of actinomycin to actinomycin monolactone and actinomycinic acid has been performed using *Actinoplanes missouriensis* in fermentation (Table 7). An inducible lactonase was isolated from this organism which catalyzes the conversion of actinomycin to its monolactone.[65] Interestingly, actinomycin monolactone has been demonstrated to occur as a biosynthetic product of actinomycin producing organisms.[153]

Microbial bioconversion of peptide antibiotics through peptide bond hydrolysis is summarized in Table 7. These conversions were performed using whole cells, cell-free extracts and purified enzymes. The transformation products found were amino acids, linear peptides, and circular peptides or various combinations of these products. In the case of the hydrolysis of bleomycin B_2, the reaction products were bleomycinic acid and agmatine.[218] The former is useful in the preparation of semisynthetic bleomycins. Inversely, the production of the novel bleomycins summarized in Table 7 has been achieved through the addition of various amines to cultures of *Streptomyces verticillus* in fermentation.[55,211] Further, peptide antibiotics have been biomodified through acylation and deacylation reactions. The substrates, transformation products,

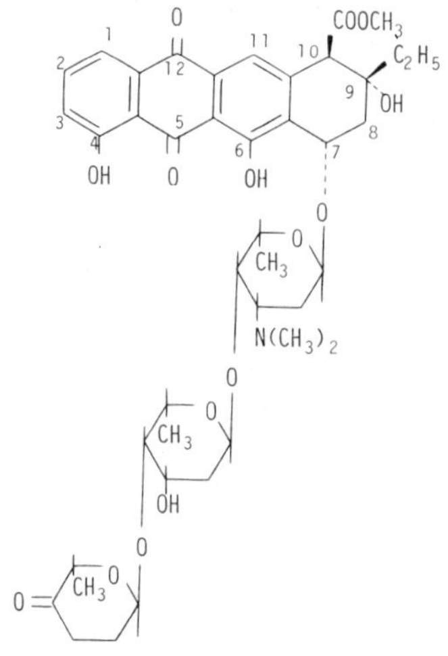

FIGURE 4. The structure of aclacinomycin A.

Table 5
BIOTRANSFORMATION OF LINCOMYCIN RELATED ANTIBIOTICS

Substrate	Transformation product	Biological agent	Ref.
1. Sulfoxidation			
Clindamycin	Clindamycin sulfoxide	*Streptomyces armentosus*	9
Lincomycin	Lincomycin sulfoxide	*Streptomyces lincolnensis*	11
2. Phosphorylation			
Celesticetin	Celesticetin phosphate	*Streptomyces* sp.	3
Clindamycin	Clindamycin-3-phosphate	*Streptomyces coelicolor*[a]	31
Lincomycin	Lincomycin-3-phosphate	*Streptomyces* sp.	3
4′-Depropyl-4′-ethyllincomycin	Phosphate of substrate	*Streptomyces* sp.	3
S-Demethyl-S-ethyllincomycin	Phosphate of substrate	*Streptomyces* sp.	3
1′-Demethyllincomycin	Phosphate of substrate	*Streptomyces* sp.	3
3. Nucleotidylation			
Clindamycin	Clindamycincytidine-5′-PO_4	*S. coelicolor*	4, 10
Clindamycin	Clindamycinadenosine-5-PO_4	*S. coelicolor*	4, 10
Clindamycin	Clindamycinuridine-5-PO_4	*S. coelicolor*	4, 10
Clindamycin	Clindamycinguanosine-5-PO_4	*S. coelicolor*	4, 10
4. Demethylation			
Celesticetin	*N*-Demethylcelesticetin	*Streptomyces caelestis*	8
Celesticetin	*N*-Demethyl-7-*O*-demethylcelesticetin	*S. caelestis*	8
Celesticetin	7-*O*-Demethylcelesticetin	*S. caelestis*	7
Clindamycin	1′-Demethylclindamycin	*Streptomyces punipalus*	9
Lincomycin	1-Demethylthio-1-hydroxylincomycin	*S. lincolnensis*	11
5. Hydroxymethylation			
N-Demethylclindamycin	*N*-Demethyl-*N*-hydroxymethylclindamycin	*S. lincolnensis*	6

Note: The biological agents were microorganisms in fermentation with the following exception. The superscript "a" indicates that a cell-free extract was used as the biological agent.

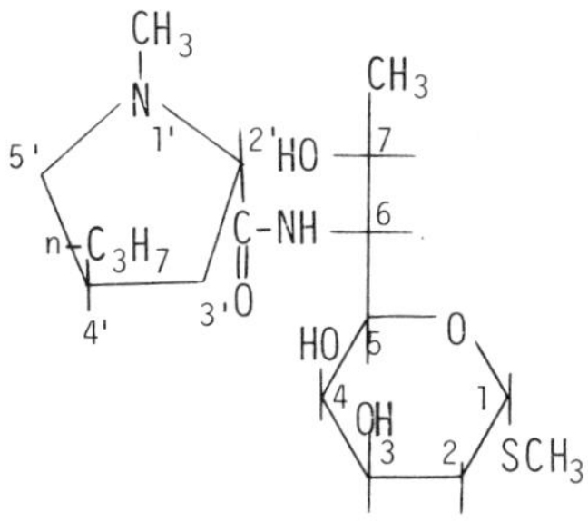

FIGURE 5. The structure of lincomycin.

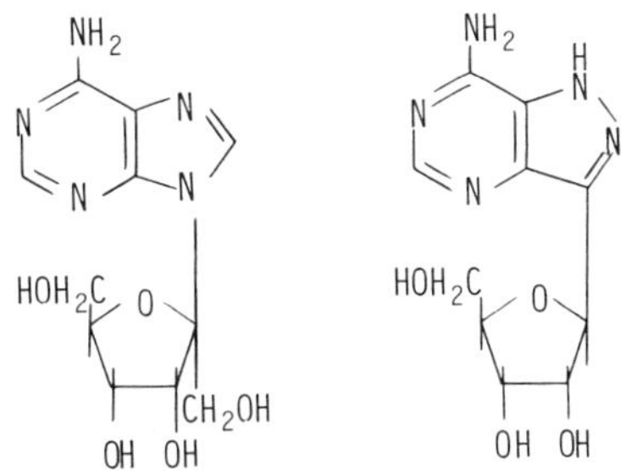

1 2

FIGURE 6. Structures of the *N*-nucleoside psicofuranine (1) and the C-nucleoside formycin (2).

and biological agents catalyzing these forms of conversion are pesented in Table 7. The structure of actinomycin D, which is numbered in the typical manner, is presented (Figure 7).

H. Biotransformation of Griseofulvin and Analogs

Modification of griseofulvin and its analogs by microbial means has been very limited. The fact that analogs can be prepared relatively easily by synthesis in greater variety and the relatively low market potential for griseofulvin analogs probably accounts for this. The structure of griseofulvin is given below.

I. Biotransformation of Chloramphenicol

Chloramphenicol is a relatively simple antibiotic whose structure is indicated below. Very few microbial transformations of this compound have been reported, and almost all of them have been some form of acylation. Both mono- and diacylation have been reported, and in the former, acylation may occur at either C-1 or C-3. Several publications discuss acetyl transferases which acylate chloramphenicol without specific mention of compound prepared.[45,167,179,181]

J. Biotransformation of Miscellaneous Antibiotics

The final group of antibiotics covered represents no unified structural family and is termed miscellaneous. Microbial biotransformations of this group have been performed through oxygenation, deacylation, acylation, esterification, hydrolysis, reduction, oxidation, isomerization, and amination. These conversions are summarized in Table 10.

Table 6
BIOTRANSFORMATION OF NUCLEOSIDE ANTIBIOTICS

Substrate	Transformation product	Biological agent	Ref.
1. Deamination			
Formycin	Formycin B	*Escherichia coli,*	216
		Nocardia interforma,[d]	217
		Streptomyces aureus NIHJ-27,[d]	217
		Streptomyces griseus NIHJ-7,	217
		Streptomyces albus NIHJ-129,[b]	217
		Streptomyces albogriseus NIHJ-130,[b]	217
		Streptomyces coelicolor NIHJ-146,[b]	217
		Streptomyces ruber NIHJ-127[b]	217
Blasticidin S	Deaminohydroxyblasticidin S	*Aspergillus terreus*[c]	243
2. Oxidation			
Formycin B	Oxoformycin	*N. interforma,*	169
		Pseudomonas fluorescenes,	169
		Streptomyces kasugaensis	169
3. Amination			
Formycin B	Formycin	*Streptomyces* sp. MA406-A-1,	139
		N. interforma,	169
		Xanthomonas oryzae	169
4. Hydrolysis			
Cordycepin	Adenine and 3-deoxy-3-aminoribose	*Bacillus subtilis*	163
Toyocamycin	Sangivamycin	*Streptomyces rimosus* (ATCC No. 14,673)[c]	197, 210
5. Isomerization			
Showdomycin	Isoshowdomycin	*Streptomyces* sp. No. 383	150
6. Phospharylation			
Formycin B	5′-*O*-Formycin phosphate	*N. interforma*	169
Tubercidin	5′-*O*-Tubercidin phosphate	*Serratia marcescens* IAM1223	189

Note: The biological agents were microorganisms in fermentation with the following exceptions. The superscripts indicate catalysis by:

[a] Whole cells and cell extracts.
[b] Cell extracts.
[c] Purified enzyme.
[d] Whole cells and purified enzyme.

Table 7
BIOTRANSFORMATION OF ACTINOMYCINS AND PEPTIDE ANTIBIOTICS

Substrate	Transformation product	Biological agent	Ref.
1. Hydrolysis			
Actinomycin D	Actinomycin monolactone and actinomycinic acid	*Actinoplanes missouriensis*[b,c]	65
Bleomycin B_2	Bleomycinic acid and agmatine	*Fusarium anguioides*[d]	218
Circulin, polymixin and tyrocidin	Consituent amino acids	*Pseudomonas*, FC-1	228
Echinomycin, etamycin, staphylomycin, stendomycin and vernamycin Bα	Complete linear peptides	*A. missouriensis*[c]	66
Gramicidin S	Linear peptide	*Bacillus subtilis*[c]	247
Polymixin E	Amino acids and peptide	*Bacillus polymyxa*[c]	235
Polymixin E	Cyclic and linear peptides	*Bacillus colistinus*[d]	69
2. Decylation			
Amphomycin	Deacylamphomycin	*Corynebacterium* sp.	229
Polymixin E	Deacylpolymixin E	*Pseudomonas*, M-6-3[a]	190
Polymixin S	Deacylpolymixin S	*Pseudomonas*, M-6-3[a]	190
3. Acylation			
Virginiamycin (M-factor)	*O*-Acetyl-M-factor	*Staphylococcus aureus*	38
4. Precursor Directed Biosynthesis			
Cis-4-methylproline and endogenous actinomycin precursor (EAP)	Actinomycin K_{1c} and actinomycin K_{2c}	*Streptomyces parvullus*	78
Trans-4-methylproline + EAP	Actinomycin 1t and actinomycin K_2t	*S. parvullus*	78
4-Methylproline and EAP	3 Novel actinomycins	*Streptomyces antibioticus*	245
DL-Pipecolic acid + EAP	Actinomycin Pip 2, actinomycin Pip 1α, and actinomycin Pip 1β	*S. antibioticus*	51
Azetidine-1-carboxylic acid + EAP	Azetomycins I, II, and III	*S. antibioticus*	52
Sarcosine and EAP	Novel actinomycins	*Streptomyces*, BOP476	170
D-Isolencine and EAP	2 Novel *N*-methylisoleucine containing actinomycins	*Streptomyces chyrsomallus*	76
Norvaline and EAP	Actinomycin AuNv	auranthin producing organism	158
Glycine and EAP	Actinomycin AuG1	auranthin producing organism	158
Amines and endogenous bleomycin precursor (EBP)	8 Novel bleomycins	*Streptomyces verticillus*	55
Exogenously supplied amines (especially diamines and triamines) and EBP	42 Bleomycins	*S. verticillus*	211

Note: The biological agents were microorganisms in fermentation with the following exceptions. The supercripts indicate catalysis by:

[a] Whole cells and cell-free extracts.
[b] Cell-free extracts.
[c] Purified enzymes.
[d] Whole cells and purified enzymes.

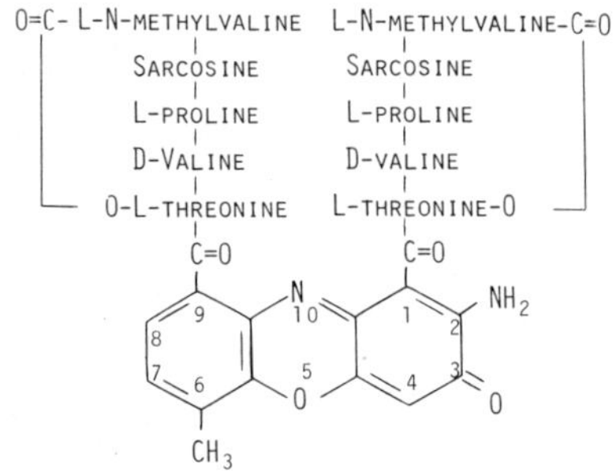

FIGURE 7. The structure of actinomycin D.

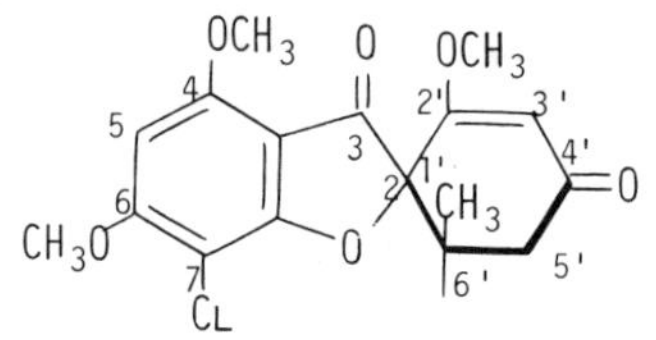

FIGURE 8. The structure of griseofulvin.

Table 8
BIOTRANSFORMATIN OF GRISEOFULVIN AND ANALOGS

Substrate	Transformation product	Biological agent	Ref.
Griseofulvin	5′-Hydroxy-griseofulvin	*Streptomyces cinerocrocatus* NRRL3443,	2
		Cephalosporium curtipes	21
	6′-Hydroxy-methylgriseofulvin	*C. curtipes*	21
	Griseofulvic acid	*Botyris allii,*	22
	6-Demethyl-griseofulvin	*Cercospora melonis,*	22
	4-Demethyl-griseofulvin	*Microsporium canis*	22
Dehydrogriseofulvin	Griseofulvin	*S. cinerocrocatus* NRRL3443	2
Griseophenone A	Dehydrogriseofulvin	*Trametes gibbose,*	146
		Fomes robustus,	146
		Trametes heteromorpha,	146
		Trametes sanguines,	146
		Coriolus fibula,	146
		Coriolus hirsutus,	146
		Gonodema lucidium,	146
		Gloseoporium lacticolor	146

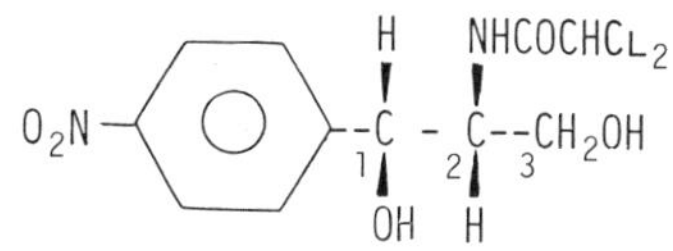

FIGURE 9. Structure of the antibiotic chloramphenicol.

Table 9
BIOTRANSFORMATION OF CHLORAMPHENICOL

Transformation product[a]	Biological agent	Ref.
1. Acylation		
1-*O*-Acetylchloramphenicol	*Streptomyces coelicolor,*	5
	Streptomyces griseus	44
Monoacetylchloramphenicol	*Staphylococcus aureus* S1174 and 1047, *S. coelicolor* Muller[b]	136
3-*O*-Acetylchloramphenicol	*S. coelicolor,*	44
	S. griseus,	
	Escherichia coli	136
	ML4079,[b]	
	E. coli strains,[b]	178
	S. aureus C 22.1,[c]	180
	E. coli $R_6$1K10,[c]	180
	E. coli K-12R5	198
3-*O*-Propionylchloramphenicol	*S. coelicolor*	5
	S. griseus	44
3-*O*-Butrylchloramphenicol	*S. griseus*	44
3-*O*-Isobutyrylchloramphenicol	*S. coelicolor,*	5
	S. griseus	44
1,3-Di-*O*-acetylchloramphenicol	*E. coli* strains,	178
	S. aureus C 22.1,[c]	180
	E. coli $R_6$1K10,[c]	180
	E. coli K-12R5[d]	198
2. Hydrolysis		
1-(*p*-Nitrophenyl)-2-aminopropan-1,3-diol	*Streptomyces* strain 135	110
1-(*p*-Nitrophenyl)-2-acetamido-propan-1,3-diol		110
1-(*p*-Nitrophenyl)-2-aminopropan-1,3-diol	*Bacillus subtilis*	195,196
	No. 0255,[d]	
	Bacillus micoides	195,196
	No. 04573,[d]	
	Proteus vulgaris	195,196
	No. 0924,[d]	
	E. coli No. 04839[d]	195,196
3. Reduction		
1-(*p*-Aminophenyl)-2-dichloroacetamidopropan-1,3-diol	*Pseudomonas aeruginosa*[c]	43
3-(*p*-Aminophenyl)serine	*B. subtilis* No. 0255,	195
	B. mycoides No. 04573,	195
	P. vulgaris No. 0924,	195
	E. coli No. 04839	195
p-Aminobenzaldehyde	*B. mycoides* No. 04573,	195
p-Aminobenzoic acid	*P. vulgaris* No. 0924,	195
	E. coli No. 04839	195

Table 9 (continued)
BIOTRANSFORMATION OF CHLORAMPHENICOL

Transformation product[e]	Biological agent	Ref.
4. Other Transformations		
α-Amino-β-hydroxy-*p*-nitropropiophenone	*E. coli* No. 04839	195
α-Dichloracetamido-β-hydroxy-*p*-nitropropiophenone	*B. subtilis* No. 0255,	195
	B. mycoides No. 04573,	195
	P. vulgaris No. 0924,	195
3-(*p*-Nitrophenyl)-serine	*E. coli* No. 04839	195
p-Nitrobenzoic acid	*Streptomyces* strain 132,	110
	B. mycoides No. 04573,	195
	E. coli No. 04839	195
p-Nitrobenzaldehyde	*B. subtilis* No. 0255,	195
	B. mycoides No. 04573,	195
	P. vulgaris No. 0924,	195
	E. coli No. 04839	195
p-Nitrobenzyl alcohol ethanolamine	*Streptomyces* strain 135,	100
	B. subtilis No. 0255,	195
	B. mycoides No. 04573,	195
	P. vulgaris No. 0924,	195
	E. coli No. 94839	195

Note: The biological agents were microorganisms in fermentation with the following exceptions. The superscripts indicate catalysis by:

[a] Whole cells and cell extracts.
[b] Cell extracts.
[c] Purified enzyme.
[d] Whole cells and purified enzyme.
[e] The substrate was chloramphenicol in all cases.

Table 10
BIOTRANSFORMATION OF MISCELLANEOUS ANTIBIOTICS

Substrate	Transformation product	Biological agent	Ref.
1. Oxygenation			
Fusidic acid	3-Oxofusidic acid	*Corynebacterium simplex*	42
Fusidic acid	6-Oxo and 7-oxofusidic acid	*Acrocylindrium oryzae*	224
Fusidic acid	11-Oxofusidic acid	*Fusidium coccineum*	224
Mycophenolic acid	4-Hydroxymycophenolic	12 fungi	71
Mycophenolic acid	3′-Oxomycophenolic acid	3 fungi	71
Mycophenolic acid	Mycophenolic acid lactol	*Streptomyces*, sp and 5 fungi	71
Mycophenolic acid	Mycophenolic acid dihydrobenzofuran	5 fungi	71
Mycophenolic acid	Mycophenolic acid hydroxylactone	3 fungi	71
Mycophenolic acid	Mycophenolic acid benzyl alcohol	*Streptomyces*, sp and 5 fungi	71
Mycophenolic acid	Mycophenolic acid dihydrobenzofuran	*Trichoderma koningi*	71
Novobiocin	Allylic methylhydroxylated novobiocins	*Sebekia benihana*	175
Cis-prophenylphosphonate	Fosfomycin	20 fungi	230
Rifamycin B	Rifamycin Y	*Streptomyces mediterranei*	100
12a-Deoxytetracycline	Tetracycline	*Curvularia lunata, Curvularia pallescens* and *Botrytis cinerea*	64a

Table 10 (continued)
BIOTRANSFORMATION OF MISCELLANEOUS ANTIBIOTICS

Substrate	Transformation product	Biological agent	Ref.
2. Deacylation			
Olivomycin A	Deisobutyrylolivomycin A	*Whetzelinia sclerotiorum*	171
Chromomycin A_2	Deisobutyrylchromomycin A_2	*W. sclerotiorum*	171
3. Acylation			
Cycloheximide	Cycloheximide acetate	*Cunninghamella blakesleeana*	67
4. Esterification			
Rifamycin S	Rifamycin L	*S. mediterranei*	98
5. Hydrolysis			
Antimycin A	Blastamycic acid	*Aspergillus ochraceus*	192
Mikamycin B	Mikamycin B acid	*Streptomyces mikakaensis*[a,c]	86,87
Novobiocin	Novenamine and 3-{3 methylbuten-2-yl}-4-hydroxybenzoic acid	nonmotile bacterium, NRRL B3652	176
6. Oxidation and Reduction			
Ascohitine	Dihydroascochitine	*Fusarium lycopersici*	145
Dihydroabikoviromycin	Abikoviromycin	*Streptomyces olivaceus*[c]	209
7. Isomerization			
Novobiocin	Isomerized novobiocin	*Actinomyces sphaeroides*[c]	14
Rifamycin S	Rifamycin G	*Nocardia mediterranei*	99
8. Amination			
Ketomycin	3-Cyclohexeneglycine	*Bacillus subtilis*[d]	84

Note: The biological agents were microorganisms in fermentation with the following exceptions. The superscripts indicate catalysis by:

[a] Whole cells and cell-free extracts.
[b] Cell-free extracts.
[c] Purified enzymes.
[d] Whole cells and purified enzymes.

REFERENCES

1. **Abraham, E. P. and Chain, E.,** An enzyme from bacteria able to destroy penicillin, *Nature (London),* 146, 837, 1940.
2. **Andres, W. W., McGahren, W. J., and Kunstmann, M. P.,** (+)-5′-Hydroxygriseofulvin, *Tetrahedron Lett.,* 3777, 1969.
3. **Argoudelis, A. D. and Coats, J. H.,** Microbial transformation of antibiotics, *J. Antibiot.,* 22, 341, 1969.
4. **Argoudelis, A. D. and Coats, J. H.,** Microbial transformation of antibiotics, *J. Am. Chem. Soc.,* 93, 534, 1971.
5. **Argoudelis, A. D. and Coats, J. H.,** Microbial transformation of antibiotics. VI. Acylation of chloramphenicol by *Streptomyces coelicolor, J. Antibiot.,* 24, 206, 1971.
6. **Argoudelis, A. D., Coats, J. H., and Magerlein, B. J.,** Microbial transformation of antibiotics, *J. Antibiot.,* 25, 191, 1972.
7. **Argoudelis, A. D., Coats, J. H., Lemaux, P. G., and Sebek, O. K.,** Antibitoics produced by mutants of *Streptomyces caelestis, J. Antibiot.,* 25, 445, 1972.
8. **Argoudelis, A. D., Coats, J. H., Lemaux, P. G., and Sebek, O. K.,** Antibiotics produced by mutants of *Streptomyces caelestis, J. Antibiot.,* 26, 7, 1973.
9. **Argoudelis, A. D., Coats, J. H., Mason, D. J., and Sebek, O. K.,** Microbial transformation of antibiotics, *J. Antibiot.,* 22, 309, 1969.
10. **Argoudelis, A. D., Coats, J. H., and Mizsak, S. A.,** Microbial transformations of antibiotics, clindamycin ribonucleotides, *J. Antibiot.,* 30, 474, 1977.
11. **Argoudelis, A. D. and Mason, D. J.,** Microbial transformation of antibiotics, *J. Antibiot.,* 22, 289, 1969.
12. **Asbell, M. A., Schwartzenbach, E., Bullock, F. J., and Yesair, D. W.,** Daunomycin and adriamycin metabolism *via* reductive glycosidic cleavage, *J. Pharmacol. Exp. Ther.,* 182, 63, 1972.

13. **Aszalos, A. A., Bachur, N. R., Hamilton, B. K., Langlykke, A. F., Roller, P. P., Sheikh, M., Sutphin, M. S., Thomas, M. C., Wareheim, D. A., and Wright, L. H.**, Microbial reduction of the side-chain carbonyl of daunorubicin and N-acetyldaunorubicin, *J. Antibiot.*, 30, 50, 1977.
14. **Batyrova, A. S. and Egorov, N. S.**, An enzymatic transformation of novobiocin and its inactive analogues in culture of *Actinomyces spheroides*, strain 35, *Antibiotiki*, 19, 1089, 1974.
15. **Beneviste, R. and Davies, J.**, Enzymatic acetylation of aminoglycoside antibiotics by *Escherchia coli* carrying an R factor, *Biochemistry*, 10, 1787, 1971.
16. **Beneviste, R. and Davies, J.**, Aminogylcoside antibiotic inactivating enzymes in *Actinomycetes* similar to those present in clinical isolates of antibiotic-resistant bacteria, *Proc. Nat. Acad. Sci. U.S.A.*, 70, 2276, 1973.
17. **Beneviste, R., Yamada, F., and Davies, J.**, Enzymatic adenylylation of streptomycin and spectinomycin by R-factor resistant *Escherchia coli*, *Infec. Immunity*, 1, 109, 1970.
18. **Beneviste, R., Yamada, F., and Davies, J.**, R-factor mediated gentamicin resistance: a new enzyme which modifies aminoglycoside antibiotics, *FEBS Lett.*, 14, 293, 1971.
19. **Biddlecome, S., Haas, M., Davies, J., Miller, G. H., Rane, D. F., and Daniels, P. J. L.**, Enzymatic modification of aminoglycoside antibiotics: a new 3-N-acetylating enzyme from a *Pseudonomas aeruginosa* isolate, *Antimicrob. Ag. Chemother.*, 9, 951, 1976.
19a. **Blumauerova, M., Jizba, J., Stajner, K., and Vanek, Z.**, Effect of DL-ethionine on the biosynthesis of anthracyclines, *Biotech. Lett.*, 471, 1979.
20. **Blumauerova, M., Kralovcova, E. Mateju, J., and Vanek, Z.**, Biotransformation of anthracyclinones in *Streptomyces coeruleorubidus* and *Streptomyces galilaeus*, *Folia Microbiol.*, 24, 117, 1979.
21. **Bod, P., Szarka, E., Gyimesi, J., Horvath, Gy., Vajna-Mehesfalvi, Z., and Horvath, L.**, Microbial transformation of griseofulvin by *Cephalosporium curtipes*, *J. Antibiot.*, 26, 101, 1973.
22. **Boothroyd, B., Napier, E. J., and Somerfield, G. A.**, The demethylation of griseofulvin by fungi, *Biochem. J.*, 80, 34, 1961.
23. **Brandl, E.**, Zur kenntnis der penicillinamidase, *Z. Physiol Chem.*, 342, 86, 1965.
24. **Brzezinska, M., Beneviste, R., Davies, J. Daniels, P. J. L., and Weinstein, J.**, Gentamicin resistance in strains of *Pseudomonas aeruginosa* mediated by enzymatic N-acetylation of the deoxystreptamine moiety, *Biochemistry*, 11, 761, 1972.
25. **Brzezinska, M. and Davies, J.**, Two enzymes which phosphorylate neomycin and kanamycin in *Escherchia coli* strains carrying R factors, *Antimicrob. Ag. Chemother.*, 3, 266, 1973.
26. **Chevereau, M., Daniels, P. J. L., Davies, J., and LeGoffic, F.**, Aminoglycoside resistance in bacteria mediated by gentamicin acetyltransferase II, an enzyme modifying the 2′-amino group of aminoglycoside antibiotics, *Biochemistry*, 13, 598, 1974.
27. **Chiang, C. and Bennett, R. E.**, Purification and properties of penicillin amidase from *Bacillus megaterium*, *J. Bacteriol.*, 93, 302, 1967.
28. **Claridge, C. A., Bush, J. A., Defuria, M. D., and Price, K. E.**, Fermentation and mutation studies with a butirosin-producing strain of *Bacillus circulans*, *Dev. Ind. Microbiol.*, 15, 101, 1974.
29. **Claridge, C. A., Gourenvitch, A., and Lein, J.**, Bacterial penicillin amidase, *Nature (London)*, 187, 237, 1960.
30. **Claridge, C. A., Luttinger, J. R. and Lein, J.**, Specificity of penicillin amidase, *Proc. Soc. Exp. Biol. Med.*, 113, 1008, 1963.
31. **Coats, J. H. and Argoudelis, A. D.**, Microbial transformation of antibiotics: Phosphorylation of clindamycin by *Streptomyces coelicolor*, Muller, *J. Bacteriol.*, 108, 459, 1971.
32. **Cole, M.**, Penicillins and other acylamino compounds synthesized by the cell-bound penicillin acylase of *Escherichia coli*, *Biochem. J.*, 115, 747, 1969.
33. **Corcoran, J. W.**, S-adenosylmethionine: erythromycin C transferase, *Methods Enzymol.*, 43, 487, 1975.
34. **Courvalin, P. and Davies, J.**, Plasmid-mediated aminoglycoside phosphotransferase of broad substrate range that phosphorylates amikamicin, *Antimicrob. Ag. Chemother.*, 11, 619, 1977.
35. **Daum, S. J., Rosi, D., and Goss, W. A.**, Production of antibiotics by biotransformation of 2,4,6/3,5-pentahydroxycyclohexanone and 2,4/3,5-tetrahydroxycyclopentanone by a deoxytetreptamine-negative mutant of *Micromonospora purpura*, *J. Am. Chem. Soc.*, 99, 283, 1977.
36. **Daum, S. J., Rosi, D., and Goss, W. A.**, Mutational biosynthesis by idiotrophs of *Micromonospora purpura*. II. Conversion of non-amino cyclitols to aminoglycoside antibiotics, *J. Antibiot.*, 30, 98, 1977.
37. **Demain, A. L., Walton, R. B., Newkirk, J. F., and Miller, I. M.**, Microbial degradation of cephalosporin C, *Nature (London)*, 199, 909, 1963.
38. **de Meester, C. and Rondelet, J.**, Microbial acetylation of M factor of virginiamycin, *J. Antibiot.*, 29, 1297, 1976.
39. **Doi, O., Kondo, S., Tanaka, N, and Umezawa, H.**, Purification and properties of kamamycin phosphorylating enzyme from *Pseudomonas aeruginosa*, *J. Antibiot.*, 22, 273, 1969.

40. **Doi, O., Ogura, M., Tanaka, N. and Umezawa, H.,** Inactivation of kanamycin, neomycin, and streptomycin by enzyme obtained in cells of *Pseudomonas aeruginosa, Appl. Microbiol.*, 16, 1276, 1968.
41. **Doi, O., Miyamoto, M., Tanaka, N, and Umezawa, H.,** Inactivation and phosphorylation of kanamycin by drug-resistant *Staphylococcus aureus, Appl. Microbiol.*, 16, 1282, 1968.
42. **Dvonch, W., Greenspan, G., and Alburn, H. E.,** Microbial oxidation of fusidic acid, *Experientia*, 22, 517, 1966.
43. **Egami, F., Ebata, M., and Sato, R.,** Reduction of chloromycetin by a cell-free bacterial extract and its relation to nitrite reduction, *Nature (London)*, 167, 118, 1951.
44. **El-Kersh, T. A. and Plourde, J. R.,** Boitransformation of antibiotics I. Acylation of chloramphenicol by spores of *Streptomyces griseus* isolated from the Egyptian soil, *J. Antibiot.*, 29, 292, 1976.
45. **El-Kersh, T. A. and Plourde, J. R.,** Biotransformation of antibiotics II. Investigation of the chloramphenicol acetyltransferase in *Streptomyces griseus, J. Antibiot.*, 29, 1189, 1976.
46. **Endo, T. and Perlman, D.,** Transglycosylation of neamine, *J. Antibiot.*, 25, 681, 1972.
47. **Endo, T. and Perlman, D.,** The transglycosylation of kan amycin A, *J. Antibiot.*, 25, 751, 1972.
48. **Feldman, L. I., Dill, I. K., Holmlund, C. E., Whaley, H. A., Patterson, E. L., and Bohonos, N.,** Microbiological transformations of macrolide antibiotics, *Antimicrob. Ag. Chemother.*, 1963, 54, 1963.
49. **Felsted, R. L., Gee, M., and Bachur, N. R.,** Rat liver daunorubicin reductase, *J. Biol. Chem.*, 249, 3672, 1974.
50. **Flickinger, M. C. and Perlman, D.,** Microbial degradation of erythromycins A and B, *J. Antibiot.*, 28, 307, 1975.
51. **Formica, J. V. and Katz, E.,** Isolation, purification and characterization of pipecolic acid-containing actiomycins, Pip 2, Pip 1α, and Pip 1β, *J. Biol. Chem.*, 248, 2066, 1973.
52. **Formica, J. V. and Apple, M. A.,** Production, isolation and properties of azetomycin, *Antimicrob. Ag. Chemother.*, 9, 214, 1976.
53. **Fugono, T., Harada, S., Higashide, E., and Kishi T.,** Studies on T-2636 antibiotics. A new component, T-2636F, *J. Antibiot.*, 24, 23, 1971.
54. **Fugono, T. Higashide, E., Suzuki, T. Yamamoto, H., Harada, S., and Kishi, T.,** Interconversion of T-2636 antibiotics produced by *Streptomyces rochei* var. *volubilis, Experientia*, 26, 26, 1970.
55. **Fujii, A., Takita, T., Shimada, N., and Umezawa, H.,** Biosyntheses of new bleomycins, *J. Antibiot.*, 27, 73, 1974.
56. **Fujisawa, Y. and Kanzaki, T.,** Role of acetyl CoA: Deacetylcephalosporin C acetyltransferase in cephalosporin C biosynthesis by *Cephalosporium acremonium, Agr. Biol. Chem.*, 39, 2043, 1975.
57. **Gatenbeck, S. and Brunsberg, U.,** Biosynthesis of penicillins, *Acta. Chem. Scand.*, 22, 1059, 1968.
58. **Haas, M., Biddlecombe, S., Davies, J., Luce, C. E., and Daniels, P. J. L.,** Enzymatic modification of aminoglycoside antibiotics: a new 6′-N-acetylating enzyme from a *Pseudomona aeruginosa* isolate, *Antimicrob. Ag. Chemother.*, 9, 945, 1976.
59. **Hamilton, B. K., Sutphin, M. S., Thomas, M. C., Wareheim, D. A., and Aszalos, A. A.,** Microbial N-acetylation of daunorubicin and daunorubicinol, *J. Antibiot.*, 30, 425, 1977.
60. **Hamilton-Miller, J. M. T.,** Penicillin acylase, *Bacteriol. Rev.*, 30, 761, 1966.
61. **Harada, S., Yamazaki, T., Hatano, K., Teuchiya, K. and Kishi, T.,** Studies on lankacidin-group (T-2636) antibiotics VII. Structure-activity relationships of lankacidin-group antibiotics, *J. Antibiot.*, 26, 647, 1973.
62. **Hennessey, T. D. and Richmond, M. H.,** The purification and some properties of β-lactamase (cephalosporinase) synthesized by *Enterobacter cloacae, Biochem. J.*, 109, 469, 1968.
63. **Higashide, E., Fugono, T., Hatano, K., and Shibata, M.,** Studies of T-2636 antibiotics. I. Taxonomy of *Streptomyces rochei* var. *volubilis* var. nov. and production of the antibiotics and an esterase, *J. Antibiot.*, 24, 1, 1971.
64. **Hiraga, K.,** Butirosin A or its derivatives, Japan Kokai Patent 76 01694, 1976, from *Chem. Abstr.*, 85, 3876, 1976.
64a. **Holmlund, C. E., Andres, W. W., and Shay, A. J.,** Microbiological hydroxylation of 12a-deoxytetracycline, *J. Amer. Chem. Soc.*, 81, 4750, 1959.
65. **Hou, C. T. and Perlman, D.,** Microbial transformations of peptide antibiotics, *J. Biol. Chem.*, 245, 1289, 1970.
66. **Hou, C. T., Perlman, D., and Schallock, M. R.,** Microbial transformation of peptide antibiotics, *J. Antibiot.*, 23, 35, 1970.
67. **Howe, R. and Moore, R. H.,** Acetylation of cycloheximide by *Cunninghamella blakesleeana, Experientia*, 24, 904, 1968.
68. **Huang, H. T., Seto, T. A., and Shull, G. M.,** Distribution and substrate specificity of benzylpenicillin acylase, *Appl. Microbiol.*, 11, 1, 1963.

69. **Ito, M., Aida, T., and Koyama, Y.,** Studies on the bacterial formation of a peptide antibiotic, colistin, *Agr. Biol. Chem.*, 30, 1112, 1966.
70. **Johnson, K. Dusart, J., Campbell, J. N., and Ghuysen, J-M.,** Exocellular β-lactamases of *Streptomyces albus* G and strains R39 and K11, *Antimicrob. Ag. Chemother.*, 3, 289, 1973.
71. **Jones, D. F., Moore, R. H., and Crawley, G. C.,** Microbial modification of mycophenolic acid, *J. Chem. Soc, (C.)*, 1970, 1725, 1970.
72. **Kabins, S., Nathan, C., and Cohen, S.,** Gentamicin-adenylyltransferase activity as a cause of gentamicin resistance in clinical isolates of *Pseudomonas aeruginosa, Antimicrob. Ag. Chemother.*, 5, 565, 1974.
73. **Kameda, Y. and Horii, S.,** The unsaturated cyclitol part of the new antibiotics, the validamycins, *J. Chem. Soc. Chem. Commun.*, 746, 1972.
74. **Kameda, Y., Horii, S., and Yamano, T.,** Microbial transformation of validamycins, *J. Antibiot.*, 28, 298, 1975.
75. **Karnetova, J., Mateju, J., Sedmera, A., Vokoun, J., and Vanek, Z.,** Microbial transformation of daunomycinone by *Streptomyces aureofaciens*, B-96, *J. Antibiot.*, 29, 1199, 1976.
76. **Katz, E., Waldron, C. R., and Meloni, M. L.,** Role of valine and isoleucine as regulators of actinomycin peptide formation by *Streptomyces chrysomallus, J. Bacteriol.*, 82, 600, 1961.
77. **Katz, E.,** Controlled biosynthesis of actinomycin, *Cancer Chemoth. Rep.*, 58, 83, 1974.
78. **Katz, E., Williams, W. K., Mason, K. T., and Mauger, A. B.,** Novel actinomycins formed by biosynthetic incorporation of cis- and trans-4-methylproline, *Antimicrob. Ag. Chemother.*, 11, 1056, 1977.
79. **Kawabe, H., Inoue, M., and Mitsuhashi, S.,** Inactivation of dihydrostreptomycin and spectinomycin by *Staphylococcus aureus, Antimicrob. Ag. Chemother.*, 5, 553, 1974.
80. **Kawabe, H., Kobayashi, F., Yamaguchi, M., Utahara, R., and Mitsuhashi, S.,** 3″-Phosphoryldihydrostreptomycin produced by the inactivating enzymes of *Pseudomonas aeruginosa, J. Antibiot.*, 24, 651, 1971.
81. **Kawabe, H., Kondo, S., Umezawa, H. and Mitsuhashi, S.,** R factor-mediated aminoglycoside antibiotic resistance in *Pseudomonas aeruginosa*. A new aminoglycoside 6′-N-acetyltransferase, *Antimicrob. Ag. Chemother.*, 7, 494, 1975.
82. **Kawabe, H., Naito, T., and Mitsuhashi, S.,** Acetylation of amikacin, a new semisynthetic antibiotic by *Pseudomonas aeruginosa* carrying an R factor, *Antimicrob. Ag. Chemother.*, 7, 50, 1975.
83. **Kaufmann, W., Bauer, K., and Offe, H. A.,** Enzymatic cleavage and resynthesis of penicillins, *Antimicrob. Ag. Ann.*, 1960, 1, 1960.
84. **Keller-Schierlein, W., Poralla, K., and Zahner, H.,** Stoffiwechselprodukte von microorganismen, *Arch. Mikrobiol.*, 67, 339, 1969.
85. **Kida, M., Igarazi, S. Okutani, T., Asako, T., Hiraga, K., and Mitsuhashi, S.,** Selective phosphorylation of the 5″-hydroxy group of ribostamycin by a new enzyme from *Pseudomonas aeruginosa, Antimicrob. Ag. Chemother.*, 5, 92, 1974.
86. **Kim, C. H., Otake, N., and Yonehara, H.,** Studies on Mikamycin B lactonase, *J. Antibiot.*, 27, 903, 1974.
87. **Kim, C. H., Endo, T., and Yonehara, H.,** Purification and some properties of mikamycin B lactonase, *Agr. Biol. Chem.*, 40, 1559, 1976.
88. **Kitao, C., Hamada, H., Ikeda, H., and Omura, S.,** Bioconversion and biosynthesis of 16-membered macrolide antibiotics XV. Final steps in the biosynthesis of leucomycin, *J. Antibiot.*, 32, 1055, 1979.
89. **Kobayashi, F., Koshi, T., Eda, J., Yoshimura, Y., and Mitsuhashi, S.,** Lividomycin resistance in *Staphylococci* by enzymatic phosphorylation, *Antimicrob. Ag. Chemother.*, 4, 1, 1973.
90. **Kobayashi, F., Yamaguchi, M., and Mitsuhashi, S.,** Inactivation of dihydrostreptomycin by *Pseudomonas aeruginosa, Japan J. Microbiol.*, 15, 381, 1971.
91. **Kobayoshi, F., Yamaguchi, M., Sato, J., and Mitsuhashi, S.,** Purification and properties of dihydrostreptomycin phosphorylating enzymes from *Pseudonomas aeruginosa, Japan J. Microbiol.*, 16, 15, 1972.
92. **Kojima, M., Ezaki, N., Amano, S., Inouye, S., and Niida, T.,** Bioconversion of ribostamycin (SF-733). II. Isolation and structure of 2-N-Acetylribostamycin, a microbiologically inactivated product of ribostamycin produced by *Streptomyces ribosidificus, J. Antibiot.*, 28, 42, 1975.
93. **Kojima, M., Inouye, S., and Niida, T.,** Bioconversion of ribostamycin (SF-733). I. Isolation and structure of 3(or 1)-N-carboxymethylribostamycin, *J. Antibiot.*, 26, 246, 1973.
94. **Kojima, M. and Satoh, A.,** Microbial semisynthesis of aminoglycosidic antibiotics by mutants of *Streptomyces ribosidificus* and *Streptomyces kanamyceticus, J. Antibiot.*, 26, 784, 1973.
95. **Komori, T., Kunugita, K., Nakahara, K. Aoki, H., and Imanaka, H.,** Production of 3-aminonocardicinic acid from nocardicin C by microbial enzymes, *Agr. Biol. Chem.*, 42, 1439, 1978.
96. **Kondo, S., Okanishi, M., Utahara, R., Maeda, K., and Umezawa, H.,** Isolation of kanamycin and paromamine inactivated by *Escherichia coli* carrying R factor, *J. Antibiot.*, 21, 29, 1968.
97. **Kuwabara, S. and Abraham, E. P.,** Some properties of two extracellular β-lactamases from *Bacillus cereus*, 569/H, *Biochem. J.*, 103, 27C, 1967.

98. **Lancini, G. C., Gallo, G. G., Startori, G., and Sensi, P.,** Isolation and structure of rifamycin L and its biogenetic relationship with other rifamycins, *J. Antibiot.*, 22, 369, 1969.
99. **Lancini, G. and Sartori, G.,** Rifamycin G, a further product of *Nocardia mediterranei* metabolism, *J. Antibiot.*, 29, 466, 1976.
100. **Lancini, G. C., Thiemann, J. E., Sartori, G., and Sensi, P.,** Biogenesis of rifamycins. The conversion of rifamycin B into rifamycin Y, *Experientia*, 23, 899, 1967.
101. **Le Goffic, F. and Martel, A.,** La resistance aux aminosides provoquée par une isoenzyme la kanamycine acetyltransferase, *Biochemie*, 56, 893, 1974.
102. **Le Goffic, F., Martel, A., Capmau, M. L., Boca, B., Goebel, P., Charden, H., Soussy, C. J., Duval, J., and Bouanchaud, D. H.,** New plasmid-mediated nucleotidylylation of aminoglycoside antibiotics in *Staphylococcus aureus*, *Antimicrob. Ag. Chemother.*, 10, 258, 1976.
103. **Le Goffic, F. Martel, A., Moreau, N., Capmau, M. L. Soussy, C. J., and Duval, J.,** 2″-O-Phosphorylation of gentamicin components by a *Staphylococcus aureus* strain carrying a plasmid, *Antimicrob. Ag. Chemother.*, 12, 26, 1977.
104. **Le Goffic, F., Martel, A., and Witchitz, J.,** 3-N Enzymatic acetylation of gentamicins, tobramycin, and kanamycin by *Escherichia coli* carrying an R factor, *Antimicrob. Ag. Chemother.*, 6, 680, 1974.
105. **Le Mahieu, R. A., Ax, H. A., Blount, J. F., Carson, M., Despraux, C. W., Prues, D. L., Scannell, J. P., Weiss, F., and Kierstead, R. W.,** A new semisynthetic macrolide antibiotic, 3-0-oleandrosyl-5-0-desosaminylerythronolide oxime, *J. Antibiot.*, 29, 728, 1976.
106. **Maeda, K., Kondo, S., Okanishi, M., Utahara, R., and Umezawa, H.,** Isolation of paromamine inactivated by *Pseudonomas aeruginosa*, *J. Antibiot.*, 21, 458, 1968.
107. **Maezawa, I., Hori, T., Kinumaki, A., and Suzuki, M.,** Biological conversion of narbonolide to picromycin, *J. Antibiot.*, 26, 771, 1973.
108. **Maezawa, I., Kinumaki, A., and Suzuki, M.,** Biological glycosidation of macrolide aglycones. I. Isolation and characterization of 5-0-mycaminosyl narbonolide, *J. Antibiot.*, 29, 1203, 1976.
109. **Majer, J., Martin, J. R., Egan, R. S., and Corcoran, J. W.,** Antibiotic glycosides. 8. Erythromycin D a new macrolide antibiotic, *J. Am. Chem. Soc.*, 99, 1620, 1977.
110. **Malik, V. S. and Vining, L. C.,** Metabolism of chloramphenicol by the producing organism, *Can. J. Microbiol.*, 16, 173, 1970.
111. **Marconi, W., Bartoli, F., Cecere, F., Galli, G., and Morisi, F.,** Synthesis of penicillins and cephalosporins by penicillin acylase entrapped in fibers, *Agr. Biol. Chem.*, 39, 277, 1975.
112. **Marshall, V. P., McGovren, J. P., Richard, F. A., Richard, R. E., and Wiley, P. F.,** Microbial metabolism of anthracycline antibiotics daunomycin and adriamycin, *J. Antibiot.*, 31, 336, 1978.
113. **Marshall, V. P., Reisender, E. A., Reineke, L. M., Johnson, J. H., and Wiley, P. F.,** Reductive microbial conversion of anthracycline antibiotics, *Biochemistry*, 15, 4139, 1976.
114. **Marshall, V. P., Reisender, E. A., and Wiley, P. F.,** Bacterial metabolism of daunomycin, *J. Antibiot.*, 29, 966, 1976.
115. **Mateju, J., Cudlin, J., Steinerova, N., Blumauerova, M., and Vanek, Z.,** Isolation of glucosyltransferase from *Streptomyces aureofaciens*, *Folia Microbiol*, 23, 337, 1978.
116. **Matsuhashi, Y., Sawa, T., Kondo, S., and Takeuchi, T.,** Aminoglycoside 3′-phosphotransferase in *Bacillus circulins* producing butirosins, *J. Antibiot.*, 30, 435, 1977.
117. **Matsuhashi, Y., Yagisawa, M., Kondo, S., Takeuchi, T., and Umezawa, H.,** Aminoglycoside 3′-phosphotransferases I and II in *Pseudomonas aeruginosa*, *J. Antibiot.*, 28, 442, 1975.
118. **McCarville, M. and Marshall, V.,** Partial purification and characterization of a bacterial enzyme catalyzing reductive cleavage of anthracycline glycosides, *Biochem. Biophys. Res. Commun.*, 74, 331, 1977.
119. **Miller, A. L. and Walker, J. B.,** Enzymatic phosphorylation of streptomycin by extracts of streptomycin-producing strains of *Streptomyces*, *J. Bacteriol.*, 99, 401, 1969.
120. **Miller, A. L. and Walker, J. B.,** Accumulation of streptomycin phosphate in cultures of producers grown on a high-phosphate medium, *J. Bacteriol.*, 104, 8, 1970.
121. **Minshew, B., Holmes, R. K., Sanford, J. B., and Baxter, C. R.,** Transferable resistance to tobramycin in *Klebsiella pneumoniae* and *Enterobacter cloacae* associated with enzymatic acetylation of tobramycin, *Antimicrob. Ag. Chemother.*, 6, 492, 1974.
122. **Mitsuhashi, S., Kobayashi, F., and Yamaguchi, M.,** Enzymatic inactivation of gentamicin C components by cell-free extract from *Pseudomonas aeruginosa*, *J. Antibiot.*, 24, 400, 1971.
123. **Naganawa, H., Kondo, S., Maeda, K., and Umezawa, H.,** Structure determinations of enzymatically phosphorylated products of aminoglycoside antibiotics by proton magnetic resonance, *J. Antibiot.*, 24, 823, 1971.
124. **Naganawa, H., Yagisawa, M., Kondo, S. Takeuchi, T., and Umezawa, H.,** The structure determination of an enzymatic inactivation product of 3,4′-dideoxykanamycin B, *J. Antibiot.*, 24, 913, 1971.
125. **Nagaoka, K. and Demain, A. L.,** Mutational biosynthesis of a new antibiotic, streptomutin A, by an idiotroph of *Streptomyces griseus*, *J. Antibiot.*, 28, 627, 1975.

126. **Nakahama, K., Harada, S., and Igarasi, J.,** Studies on lankacidin-group (T-2636) antibiotics. X. Microbial conversion of lankacidin-group antibiotics, *J. Antibiot.,* 28, 390, 1975.
127. **Nakahama, K., Isawa, M., Muroi, M., Kishi, T., Uchida, M., and Igarasi, S.,** Microbial conversion of antibiotics. I. Deacylation of maridomycin by bacteria, *J. Antibiot.,* 27, 425, 1974.
128. **Nakahama, K., Kishi, T., and Igarasi, S.,** Microbial conversion of antibiotics. III. Hydroxylation of maridomycin I and josamycin, *J. Antibiot.,* 27, 433, 1974.
129. **Nakahama, K., Kishi, T., and Igarasi, J.,** Microbial conversion of antibiotics. II. Deacylation of maridomycin by *Actinomycetes, J. Antibiot.,* 27, 487, 1974.
130. **Nara, T., Okachi, R., and Misawa, M.,** Enzymatic synthesis of D(−)-α-aminobenzylpenicillin by *Kluyvera citrophila, J. Antibiot.,* 24, 321, 1971.
131. **Niida, T., Inoue, S., Shomura, J., Omato, S., Watanabe, H., Nomiya, B., and Koeda, T.,** Biological production of 9-acyl-4″-depropionyl-SF-837 compounds. Japan Kokai Patent 73 72389, 1973, from *Chem. Abstr.,* 80, 35826c, 1974.
132. **Nimi, O., Ito, G., Ohata, Y., Furayama, S., and Nomi, R.,** Streptomycin phosphorylating enzyme produced by *Streptomyces griseus, Agr. Biol. Chem.,* 35, 856, 1971.
133. **Nimi, O., Ito, G., Sueda, S., and Nomi, R.,** Phosphorylation of streptomycin at C_6-OH of streptidine moiety by an intracellular enzyme of *Streptomyces griseus, Agr. Biol. Chem.,* 35, 848, 1971.
134. **Ninet, L., Florent, J., Lunel, J., Renaut, J., Abraham, A., Lombardi, B., and Tissier, R.,** Daunorubicinol preparation by biochemical reduction of daunorubicin, Abstracts, 17.06, 5th International Fermentation Symp., Berlin, 1976.
135. **Nishiura, ., Kawada, Y., Shiomi, Y., O'ltara, K., and Kono, M.,** Microbial degradation of cephalothin by cephalothin-susceptible *Escherichia coli, Antimicrob. Ag. Chemother.,* 13, 1036, 1978.
136. **Nokano, H., Matsuhashi, Y., Takeuchi, T., and Umezawa, H.,** Distribution of chloramphenicol acetyltransferase and chloramphenicol-3-acetate esterase among *Streptomyces* and *Corynebacterium, J. Antibiot.,* 30, 76, 1977.
137. **Nomi, R., Nimi, O., and Kado, T.,** Biosynthesis of streptomycin. Part IV. Accumulation of a streptomycin precursor in the culture broth and partial purification of the precursor, *Agr. Biol. Chem.,* 32, 1256, 1968.
138. **Nomi, R., Nimi, O., and Kado, T.,** Biosynthesis of streptomycin. Part V. Purification and properties of a streptomycin precursor, *Agr. Biol. Chem.,* 32, 1454, 1969.
139. **Ochi, K., Yoshima, S., and Eguchi, Y.,** Biosynthesis of formycin. Formation of formycin from formycin B, *J. Antibiot.,* 28, 965, 1975.
140. **O'Hara, K., Kono, M., and Mitsuhashi, S.,** Structure of enzymatically acetylated sisomicin by *Pseudomonas aeruginosa, J. Antibiot.,* 27, 349, 1974.
141. **O'Hara, K., Kono, M., and Matsuhashi, S.,** Enzymatic inactivation of a new aminoglycoside antibiotic, sisomicin, by resistant strains of *Pseudomonas aeruginosa, Antimicrob. Ag. Chemother.,* 5, 558, 1974.
142. **Okachi, R. and Nara, T.,** Penicillin acylase of *Pseudomonas melanogenum, Agr. Biol. Chem.,* 37, 2797, 1973.
143. **Okanishi, M., Kondo, S., Utahara, R., and Umezawa, H.,** Phospharylation and inactivation of aminoglycosidic antibiotics by *Escherichia coli* carrying R factor, *J. Antibiot.,* 21, 13, 1968.
144. **Oki, T., Komiyama, T., Tone, H., Inui, T., Takeuchi, T., and Umezawa, H.,** Reductive cleavage of anthracycline glycosides by microsomal NADPH cytochrome C reductase, *J. Antibiot.,* 30, 613, 1977.
145. **Oku, H. and Nakanishi, T.,** Reductive detoxification of an antibiotic, ascochitine, by an insensitive fungus, *Fusarium lycopersici,* Sacc., *Naturwissenschaften,* 51, 538, 1964.
146. **Okuda, S., Isaka, H., Iida, M., Minemura, Y., Iizuka, H., and Isuda, K.,** Microbial transformation of griseophenone A, *J. Pharm. Soc. Jpn.,* 87, 1003, 1967.
147. **Okutani, T., Asako, T., Yoshioka, K., Hirago, K., and Kida, M.,** Conversion of aminoglycosidic antibiotics. Novel and efficient approaches to 3′-deoxyaminoglycosides via 3′-phosphamyl esters, *J. Am. Chem. Soc.,* 99, 1278, 1977.
148. **Omura, S., Miyazawa, J., Takeshima, H., Kitao, C., and Aizawa, .,** Induction of the bioconversion of leucomycins by glucose in a producing strain, *J. Antibiot.,* 30, 192, 1977.
149. **Omura, S., Miyazawa, J., Takeshima, H., Kitao, C., Atsumi, K., and Aizawa, M.,** Bioconversion of leucomycins and its regulation by butyrate in a producing strain, *J. Antibiot.,* 29, 1131, 1976.
150. **Ozaki, M., Kariya, T., Kato, H., and Kimura, T.,** Microbial transformation of antibiotics. I. Isomerization of showdomycin by *Streptomyces* species, *Agr. Biol. Chem.,* 36, 451, 1972.
151. **Ozanne, B., Beneviste, R., Tipper, D., and Davies, J.,** Aminoglycoside antibiotics: inactivation by phosphorylation in *Escherichia coli, J. Bacteriol.,* 100, 1144, 1969.
151a. **Paulik, R. C., Casey, M. L., and Whitlock, H. W.,** A ^{13}C-nuclear magnetic resonance study of the biosynthesis of daunomycin from $^{13}CH_3{}^{13}CO_2Na$, *J. Amer. Chem. Soc.,* 98, 3370, 1976.
152. **Perlman, D.,** Microbial transformations of antibiotics, *Process Biochem.,* 1971, 13, 1971.

153. **Perlman, D.**, Actinomycin monolactones: origin and importance, *Cancer Chemother. Rep.*, 58, 93, 1974.
154. **Perlman, D. and Langlykke, A. F.**, Occurrence of mannosidostreptomycinase, *J. Amer. Chem. Soc.*, 70, 3968, 1948.
155. **Perlman, D. and Sebek, O. K.**, Microbial transformations of antibiotics, *Pure Appl. Chem.*, 28, 637, 1971.
156. **Pollock, M. R.**, The measurement of the liberation of penicillinase from *Bacillus subtilis*, *J. Gen. Microbiol.*, 26, 239, 1961.
157. **Pollock, M. R.**, Purification and properties of penicillinases from two strains of *Bacillus licheniformis:* a chemical, physiochemical comparison, *Biochem. J.*, 94, 666, 1965.
158. **Prokopenko, V. M., Belyaeva, I. D., and Nefelova, M. V.**, Biological properties of modified actinomycins synthesized by auranthin-producing organism, *Antibiotiki*, 23, 99, 1978.
159. **Richmond, M. H.**, Wild-type variants of exopenicillinase from *Staphylococcus aureus*, *Biochem. J.*, 94, 584, 1965.
159a. **Richmond, M. H.**, β-Lactamase *(Escherichia coli* R^{+}_{TEM}), *Methods in Enzymology*, 43, 672, 1975.
160. **Rinehart, Jr, K. L.**, *Neomycin and Related Antibiotics*, John Wiley & Sons, New York, 1964, 93.
161. **Rolinson, G. N., Batchelor, F. R., Butterworth, D., Cameron-Wood, J., Cole, M., Eustace, G. C., Hart, M. V., Richards, M., and Chain, E. B.**, Formation of 6-aminopenicillanic acid from penicillin by enzymatic hydrolysis, *Nature (London)*, 187, 236, 1960.
162. **Rosi, D., Goss, W. A., and Daum, S. J.**, Mutational biosynthesis by idiotrophs of *Micromonospora purpura*. I. Conversion of aminocyclitols to new aminoglycoside antibiotics, *J. Antibiot.*, 30, 88, 1977.
163. **Rottman, F. and Guarino, A. J.**, Studies on the inhibition of *Bacillus subtilis* growth by cordyceptin, *Biochem. Biophys. Acta*, 80, 632, 1964.
164. **Rueckert, P. W., Wiley, P. F., McGovren, J. P., and Marshall, V. P.**, Mammalian and microbial cell-free conversion of anthracycline antibiotics and analogs, *J. Antibiot.*, 32, 141, 1979.
165. **Sakaguchi, K. and Murao, S.**, A new enzyme, penicillin amidase, *J. Agr. Chem. Soc. (Japan)*, 23, 411, 1950.
166. **Sakaguchi, Y., Takaishi, N., and Hachimori, A.**, Extraction and purification of kanamycin inactivation enzyme of *Escherichia coli* K 12 ML 1629, *J. Antibiot.*, 27, 248, 1974.
167. **Sands, L. C. and Shaw, W. V.**, Mechanism of chloramphenicol resistance in *Staphylococci:* characterization and hybridization of variants of chloramphemicol acetyltransferase, *Antimicrob. Ag. Chemother.*, 3, 299, 1973.
168. **Savidge, T. A. and Cole, M.**, Penicillin acylase (Bacterial), *Methods Enzymol.*, 43, 705, 1975.
169. **Sawa, T., Fukagawa, Y., Homma, I., Wakashiro, T., Takeuchi, T., and Hori, M.**, Metabolic conversion of formycin B to formycin A and to oxoformycin B in *Nocardia interforma*, *J. Antibiot.*, 21, 334, 1968.
170. **Schmidt-Kastner, G.**, The production of actinomycins by controlled biosynthesis: the F-actinomycins, *Ann. N.Y. Acad. Sci.*, 89, 299, 1960.
171. **Schmitz, H. and Claridge, C. A.**, Biotransformation of antitumor agents by a strain of *Whetzelinia sclerotiorum*, *J. Antibiot.*, 30, 635, 1977.
172. **Sebek, O. K.**, Microbial conversion of antibiotics, *Lloydia*, 37, 115, 1974.
173. **Sebek, O. K.**, The scope and potential of antibiotic conversion by microorganisms, *Acta. Microbiol. Acad. Sci. Hung.*, 22, 381, 1975.
174. **Sebek, O. K.**, Biotransformation of steroid hormones and antibiotics, in *Biotechnological Applications of Proteins and Enzymes*, Bohak, Z. and Sharon, N., Eds., Academic Press, New York, 1977, 203.
175. **Sebek, O. K. and Dolak, L. A.**, Allylic methylhydroxylated novobiocins, U.S. Patent 4, 128, 563, 1978.
176. **Sebek, O. K. and Hoeksema, H.**, Enzymic formation of novenamine, a novosylaminocoumarin moiety of novobiocin, *J. Antibiot.*, 25, 434, 1972.
177. **Sebek, O. K. and Perlman, D.**, Microbial transformation of antibiotics, *Adv. Appl. Microbiol.*, 14, 123, 1971.
178. **Shaw, W. V.**, The enzymatic acetylation of chloramphenicol by extracts of R factor-resistant *Escherichia coli*, *J. Biol. Chem.*, 242, 687, 1967.
179. **Shaw, W. V.**, Choramphenicol acetyltransferase from chloramphenicol-resistant bacteria, *Methods Enzymol.*, 43, 737, 1975.
180. **Shaw, W. V. and Brodsky, R. F.**, Characterization of chloramphenicol acetyltransferase from chloramphenicol-resistant, Staphylococcus aureus, *J. Bacteriol.*, 95, 28, 1968.
181. **Shaw, W. V., Sands, L. C., and Datta, N.**, Hybridization of variants of chloramphenicol acetyltransferase specified by fi+ and fi− R factors, *Natl. Acad. Sci. U.S.*, 69, 3049, 1972.

182. **Shibata, M. and Uyeda, M.,** Microbial transformations of antibiotics, in *Annual Reports on Fermentation Processes,* Vol. 2, Perlman, D. and Tsao, G. T., Eds., Academic Press, New York, 1978, 267.
183. **Shibata, M., Uyeda, M., and Mori, S.,** Microbial transformation of antibiotics. I. Isolation and characterization of the transformation products of maridomycin III, *J. Antibiot.,* 28, 434, 1975.
184. **Shibata, M., Uyeda, M., and Mori, S.,** Microbial transformation of antibiotics. II. Additional transformation products of maridomycin III, *J. Antibiot.,* 29, 824, 1976.
185. **Shier, W. T., Rinehart, K. L., Jr., and Gottlieb, D.,** Preparation of four new antibiotics from a mutant of *Streptomyces fradiae, Proc. Natl. Acad. Sci. U.S.,* 63, 198, 1969.
186. **Shier, W. T., Rinehart, K. L., Jr., and Gottlieb, D.,** Antibiotics containing the aminocyclitol subunit, U.S. Patent 3, 669, 838, 1972, from *Chem. Abstr.,* 77, 154584x, 1972.
187. **Shimizu, M., Masuike, T., Fujita, H., Kimura, K., Okachi, R., and Nara, T.,** Search for microorganisms producing cephalosporin acylase and enzymatic synthesis of cephalosporins, *Agr. Biol. Chem.,* 39, 1225, 1975.
188. **Shimizu, M., Masuike, T., Fujita, H., Iida, T., Kimura, K., and Nara, T.,** Characterization of 7-[1-(1H)-tetrazolylactamido]-desacetoxycephal osporanic acid enzymatically synthesized by *Kluyvera citrophila, Agr. Biol. Chem.,* 39, 1745, 1975.
189. **Shirato, S., Hoshida, K., and Miyazaki, Y.,** Antibiotic base analog, tubercidin III, Biochemical synthesis of tubercidin 5′-monophosphate, *J. Ferment. Technol.,* 46, 233, 1968.
190. **Shoji, J., Kato, T., and Hinoo, H.,** The structure of polymixin S, *J. Antibiot.,* 30, 1035, 1977.
191. **Shomura, T., Inoue, S., Ezaki, N., Tsuruoka, T., Amano, T., Watanabe, H., and Niida, T.,** Deacylation of leucomycin and antibiotic SF-837, Japan Kokai Patent 73 29148, 1973, from *Chem. Abstr.,* 80, 144395g, 1974.
192. **Singh, K. and Rakhit, S.,** Transformations of antibiotics, *J. Antibiot.,* 24, 704, 1971.
193. **Smith, A. L. and Smith, D. H.,** Gentamicin: adenine mononucleotide transferase: partial purification, characterization, and use in the clinical quantitation of gentamicin, *J. Infect. Diseases,* 129, 391, 1974.
194. **Smith, G. N. and Worrel, C. S.,** Enzymatic reduction of chloramphenicol (chloromycetin), *Biochemistry,* 24, 216, 1949.
195. **Smith, G. N. and Worrel, C. S.,** The decomposition of chloromycetin (chloramphenicol) by microorganisms, *Arch. Biochem.,* 28, 232, 1950.
196. **Smith, G. N., Worrel, C. S., and Lilligren, B. S.,** The enzymatic hydrolysis of chloramphenicol (chloromycetin), *Science,* 110, 297, 1949.
197. **Suhadolnik, R. J.,** *Nucleoside Antibiotics,* Wiley-Interscience, New York, 1970, 327.
198. **Susuki, Y. and Okamato, S.,** The enzymatic acetylation of chloramphenicol by the multiple drug-resistant *Escherichia coli* carrying R factor, *J. Biol. Chem.,* 242, 4722, 1967.
199. **Suzuki, Y., Okamato, S., and Kono, M.,** Basis of chloramphenicol resistance in naturally isolated resistant *Staphylococci, J. Bacteriol.,* 92, 798, 1966.
200. **Suzuki, M., Takamaki, T., Miyagawa, K., Ono, H., Higashide, E., and Ochida, M.,** Interconversion among leucomycin A_3, carbomycin A, carbomycin B, and maridomycin II, *Agric. Biol. Chem.,* 41, 419, 1977.
201. **Sykes, R. B. and Richmond, M. H.,** Intergeneric transfer of a β-lactamase gene between *Ps. aeruginosa* and *Escherichia coli, Nature (London),* 226, 952, 1970.
202. **Takahashi, T., Yamazaki, Y., Kato, K., and Isono, M.,** Enzymatic synthesis of cephalosporins, *J. Am. Chem. Soc.,* 94, 4035, 1972.
203. **Takasawa, S., Utahara, R., Okanishi, M., Maeda, K., and Umezawa, H.,** Studies on adenylylstreptomycin, a product of streptomycin inactivation by *Escherichia coli* carrying the R factor, *J. Antibiot.,* 21, 477, 1968.
204. **Taylor, H. D., and Schmitz, H.,** Antibiotics derived from a mutant of *Bacillus circulans, J. Antibiot.,* 29, 532, 1976.
205. **Testa, R. L. and Tilley, B. C.,** Biotransformation, a new approach to aminoglycoside biosynthesis. I. Sisomicin, *J. Antibiot.,* 28, 573, 1975.
206. **Testa, R. L. and Tilley, B. C.,** Biotransformation, a new approach to aminoglycoside biosynthesis. II. Gentamicin, *J. Antibiot.,* 29, 140, 1976.
207. **Testa, R. L., Wagman, G. H., Daniels, P. J. L., and Weinstein, M. J.,** Mutamicins biosynthetically created new sisomicin analogues, *J. Antibiot.,* 27, 917, 1974.
208. **Theriaudt, R.,** Mycarosyl macrolide antibiotics by fermentation, U.S. Patent 2,784,447, 1974, from *Chem. Abstr.,* 80, 81043t, 1974.
209. **Tsuruoka, T., Shomura, T., Ogawa, Y., Ezoki, N., Watanabe, H., Amano, S., Inouye, S., and Niida, T.,** Some properties of SF-973B substance, the enzyme catalyzing the conversion of dihydroabikoviromycin to abikoviromycin, *J. Antibiot.,* 26, 168, 1973.
210. **Uematsu, T. and Suhadolnik, R. J.,** In vivo and enzymatic conversion of toyocamycin to sangivamycin by *Streptomyces rimosus, Arch. Biochem. Biophys.,* 162, 614, 1974.

211. **Umezawa, H.,** Natural and artificial bleomycins: chemistry and antitumor activities, *Pure Appl. Chem.*, 28, 665, 1971.
212. **Umezawa, H.,** Biochemical mechanisms of resistance to aminoglycoside antibiotics, *Adv. Carbohyd. Chem. Biochem.*, 30, 183, 1974.
213. **Umezawa, H., Doi, O., Ogura, M., Kondo, S., and Tanaka, N.,** Phosphorylation and inactivation of kanamycin by *Pseudomonas aeruginosa, J. Antibiot.*, 21, 154, 1968.
214. **Umezawa, H., Okanishi, M., Kondo, S., Hamana, K., Utahara, R., and Maeda, K.,** Phosphlorylative inactivation of aminoglycosidic antibiotics by *Escherichia coli* carrying R factor, *Science*, 157, 1559, 1967.
215. **Umezawa, H., Okanishi, M., Utahara, R., Maeda, K., and Kondo, S.,** Isolation and structure of kanamycin inactivated by a cell free system of kanamycin-resistant *Escherichia coli, J. Antibiot.*, 20, 136, 1967.
216. **Umezawa, H., Sawa, T., Fukagama, Y., Homma, I., Ishizuka, M., and Takeuchi, T.,** Studies on formycin and formycin B in cells of Ehrlich carcinoma and *Escherichia coli., J. Antibiot.*, 20, 308, 1967.
217. **Umezawa, H., Sawa, T., Fukagawa, Y., Koyama, G., Murase, M., Hamada, M., and Takeuchi, T.,** Transformation of formycin to formycin B and their biological activities, *J. Antibiot.*, 18, 178, 1965.
218. **Umezawa, H., Takahashi, Y., Fujii, A., Saino, T., Shirai, T., and Takita, T.,** Preparation of bleomycinic acid: hydrolysis of bleomycin B_2 by a *Fusarium* acylagmatine amidohydrolase, *J. Antibiot.*, 26, 117, 1973.
219. **Umezawa, H., Tapasawa, S., Okanishi, M., and Utahara, R.,** Adenylylstreptomycin, a product of streptomycin inactivated by *Escherichia coli* carrying R factor, *J. Antibiot.*, 21, 81, 1968.
220. **Umezawa, H., Yagisawa, M., Matsuhashi, Y., Naganawa, H., Yamamoto, H., Kondo, S., and Takeuchi, T.,** Gentamicin acetyltransferase in *Escherichia coli* carrying R factor., *J. Antibiot.*, 26, 612, 1973.
221. **Umezawa, Y., Yagisawa, M., Sawa, T., Takeuchi, T., and Umezawa, H.,** Aminoglycoside 3′-phosphatotransferase III, a new phosphotransferase, *J. Antiobiot.*, 28, 845, 1975.
222. **Uyeda, M., Mori, S., Morita, M., Ogata, T., Mori, M., and Shibata, M.,** Microbial transformation of antibiotics III. Reacylation of 4″-depropionyl maridomycin III into maridomycin V (Maridomycin K) by *Streptomyces* sp. strain no. K-342, *J. Antibiot.*, 30, 1130, 1977.
223. **Vanek, Z., Tax, J., Komersova, I., and Echardt, K.,** Glycosidation of ε-pyrromycinone using the strain *Streptomyces galilaeus*, JA3043, *Folia Microbiol.*, 18, 524, 1973.
224. **von Daehne, W., Lorch, H., and Godtfredsen, W. O.,** Microbial transformations of fusidane-type antibiotics. A correlation between fusidic acid and hevonic acid, *Tetrahedron Lett.*, 47, 4843, 1968.
225. **Walker, M. S. and Walker, J. B.,** Streptomycin biosynthesis and metabolism. Enzymatic phosphorylation of dihydrostreptomycin-(streptidino) phosphate and dihydrostreptomycin by *Streptomyces* extracts, *J. Biol. Chem.*, 245, 6683, 1970.
226. **Walker, J. B. and Skorvaga, M.,** Phosphorylation of streptomycin and dihydrostreptomycin by *Streptomyces, J. Biol. Chem.*, 248, 2435, 1973.
227. **Walton, R. B.,** Search for microorganisms producing cephalosporin C amidase, *Develop. Ind. Microbiol.*, 5, 349, 1964.
228. **Warren, R. A. J. and Neilands, J. B.,** Microbial degradation of ferrichrome compounds, *J. Gen. Microbiol.*, 35, 459, 1964.
229. **Weber, J. M. and Perlman, D.,** Amphomycin acylase, *J. Antibiot.*, 31, 373, 1978.
230. **White, R. F., Birnbaum, J., Meyer, R. T., ten Broeke, J., Chemerda, J. M., and Demain, A. L.,** Microbial expoxidation of cis-propenylphosphonic acid, *Appl. Microbiol.*, 22, 55, 1971.
230a. **Wiley, P. F., Elrod, D. W., and Marshall, V. P.,** Biosynthesis of the anthracycline antibiotics nogalamycin and steffimycin B, *J. Org. Chem.*, 43, 3467, 1978.
231. **Wiley, P. F. and Marshall, V. P.,** Microbial conversion of anthracycline antibiotics, *J. Antibiot.*, 28, 838, 1975.
232. **Wiley, P. F., Koert, J. M., Elrod, D. W., Reisender, E. A., and Marshall, V. P.,** Bacterial metabolism of anthracycline antibiotics. Steffimycinone and steffimycinol conversions, *J. Antibiot.*, 30, 649, 1977.
233. **Wiley, P. F., Slavicek, J. M., Elrod, D. W., and Marshall, V. P.,** Microbial conversion of steffimycin to 10-dihydrosteffimycin and steffimycin B to 10-dihydrosteffimycin B, *J. Antibiot.*, 33, 819, 1980.
234. **Williams, J. W. and Northrop, D. B.,** Purification and properties of gentamicin acetyltransferase I, *Biochemistry*, 15, 125, 1976.
235. **Woyczikowska, B., Pass, L., Girdwoyn, M., and Raczynska-Bojanowska, K.,** Proteolytic degradation of polymixins by the enzymes of *Bacillus polymyxa, Acta. Biochem. Pol.*, 20, 285, 1973.
236. **Yagisawa, M., Kondo, S., Takeuchi, T. and Umezawa, H.,** Aminoglycoside 6′-N-acetyltransferase of *Pseudomonas aeruginosa:* structural requirements of substrate, *J. Antibiot.*, 28, 486, 1975.

237. **Yagisawa, M., Naganawa, H., Kondo, S., Hamada, M., Takeuchi, T., and Umezawa, H.,** Adenylyldideoxykanamycin B, a product of inactivation of dideoxykanamycin B by *Escherichia coli* carrying R factor, *J. Antibiot.*, 24, 911, 1971.
238. **Yagisawa, M., Naganawa, H., Kondo, S., Takeuchi, T., and Umezawa, H.,** Inactivation of 3′, 4′-dideoxykanamycin B by an enzyme solution of resistant *Escherichia coli* and isolation of 3′, 4′-dideoxykanamycin B 2″-guanylate and 2″-inosate, *J. Antibiot.*, 25, 492, 1972.
239. **Yagisawa, M., Naganawa, H., Kondo, S., Takeuchi, T., and Umezawa, H.,** 6′-N-acetylation of 3′,4′-dideoxykanamycin B by an enzyme in a resistant strain of *Pseudomonas aeruginosa, J. Antibiot.*, 25, 495, 1972.
240. **Yagisawa, M., Yamamoto, H., Naganawa, H., Kondo, S., Takeuchi, T., and Umezawa, H.,** A new enzyme in *Escherichia coli* carrying R-factor phosphorylating 3′-hydroxyl of butirosin A, kanamycin, neamine, and ribostamycin, *J. Antibiot.*, 25, 748, 1972.
241. **Yamada, T., Tipper, D., and Davies, J.,** Enzymatic inactivation of streptomycin by R factor-resistant *Escherichia coli, Nature (London)*, 219, 288, 1968.
242. **Yamaguchi, M., Mitsuhashi, S., Kobayashi, F., and Zenda, H.,** A 2′-N-acetylating enzyme of aminoglycosides, *J. Antibiot.*, 27, 507, 1974.
243. **Yamaguchi, J., Shibata, H., Seto, H., and Misato, T.,** Isolation and purification of blasticidin S deaminase from *Aspergillus terreus, J. Antibiot.*, 28, 7, 1975.
244. **Yamamoto, H., Yagisawa, M., Naganawa, H., Kondo, S., Takeuchi, T., and Umezawa, H.,** Kanamycin 6′-acetate and ribostamycin 6′-acetate, enzymatically inactivated by *Pseudomonas aeruginosa, J. Antibiot.*, 25, 746, 1972.
245. **Yoshida, T., Mauger, A., Weissbach, H., and Katz, E.,** Effect of structural and stereochemical methylproline isomers on actimonycin biosynthesis, *J. Bacteriol.*, 95, 952, 1968.
246. **Yoshimoto, A., Ogasawara, T., Kitamura, I., Toshikazu, O., Inui, T., Takeuchi, T., and Umezawa, H.,** Enzymatic conversion of aclacinomycin A to Y by a specific oxidoreductase in *Streptomyces, J. Antibiot.*, 32, 472, 1979.
247. **Yukioka, M., Saito, Y., and Otani, S.,** Enzymatic hydrolysis of gramicidin S, *J. Biochem.*, 60, 295, 1966.

Chapter 4

MICROBIAL TRANSFORMATION OF IMPORTANT INDUSTRIAL HYDROCARBONS

Ching T. Hou

TABLE OF CONTENTS

I. INTRODUCTION

Microbial transformations are defined, in this article, as microbial (or enzymatic) reactions which produce a chemical in which most of the carbon skelton is derived intact from the substrate. Useful transformations may be simple one-step reactions or those involving a few sequential steps, where reaction products can be isolated and identified. The subjects of biogenesis or fermentation in which products are found, regardless of the growth substrates, will be excluded from this article.

Microbial transformation of nonsteroidal compounds have been described and tabulated.[1] Transformation and biosynthesis by hydrocarbon-degrading microorganisms have been discussed extensively.[2] Chemical oxidations involving microorganisms have been discussed,[3-5] and microbial transformations of organic compounds have been reviewed.[6,7] In addition, a number of papers have been published within the past decade, particularly in the last 5 years, which deal with industrial chemicals. The criteria for substrates and products which render industrial chemicals attractive for use are their abundance in nature, relative purity, and low cost.

Microbial transformations may be brought about by constitutive or inducible enzymes, or they may be due to the broad substrate specificity of essential enzymes in microorganisms. Ever since the fundamental work of microbiology laid down by Pasteur, microbial transformations have been largely confined to reactions involving carbohydrates and steroids. Progress in biotechnology, such as immobilization of cells or enzymes and improvements in analytical assay techniques in this decade has expanded the application of microbial transformations from the traditional transformation of steroids to include many useful industrial chemicals. From a purely scientific point of view, almost all chemicals can be transformed microbiologically or enzymatically into many kinds of products. However, from an economic point of view, microbial transformation of fine chemicals stands the best immediate chance for industrial exploitation. Since the energy crisis in 1973, it has become obvious that nonrenewable petroleum fractions should not be burned merely for their energy (the best technology available now converts less than 30% of the total energy content into useful energy). The best use of the nonrenewable petroleum fractions would convert them into useful substances by either microbial or chemical processes. Advantages of microbial transformations over chemical processes may be summarized as follows:

1. Microbial transformation takes place under mild conditions. Therefore, chemicals can be converted into desirable products without decomposition. With mild reaction conditions, a microbial process saves not only operation energy, but also capital investment.
2. A microbial process can produce stereospecific products by attacking a specific site of the substrate molecule.
3. Several reactions can be combined into one transformation step and may actually be programmed to occur in a specific sequence.

This review attempts to assemble up-to-date information on microbial transformation of hydrocarbons to useful or potentially useful industrial chemicals.

II. ALIPHATIC HYDROCARBONS

A. Alkanes

The first hydrocarbon-utilizing microorganism, *Botrytis cinerea,* was reported by Miyoshi[8] in 1895. This was followed by Sohngen in 1906 who isolated a methane utilizer, *Pseudomonas methanica.* Since then, a wide range of exploration has been directed toward the potential for growth on or transformation of hydrocarbons by microbes. The results of this exploration have been presented and analyzed in many reviews.[2,9-19]

Oxidation of alkanes occurs either at the methyl group, forming a primary alcohol, or at the subterminal position, forming a secondary alcohol. Diterminal oxidation may also occur when one methyl group is oxidized to the acid level followed by oxidation of the remaining methyl group to form the ω-hydroxy acid.

1. Methane

Methane is the main ingredient of natural gas and oil field flare gas and it is the by-product of several industrial processes. Microbial transformation of methane gas into liquid chemicals such as methanol, formaldehyde, and formic acid, or into single cell protein has high industrial potential. Methane is oxidized by a number of specified genera of bacteria (methylotrophs) according to the following scheme:

$$\begin{array}{c} \text{cell constituents} \\ \uparrow \\ CH_4 \rightarrow CH_3OH \rightarrow HCHO \rightarrow HCOOH \rightarrow CO_2 + H_2O \end{array}$$

Whittenbury et al.[20] first systematically classified obligate methane utilizers into two groups based on their internal membrane structure type. Type I obligate methane utilizers have their membrane arranged in bundles of vesicular disks and form cysts at their resting stage. These organisms utilize the ribulose monophosphate carbon assimilation pathway and have an incomplete TCA cycle. Type II obligate methane utilizers have their membranes arranged in pairs around cell periphery and form exospores or lipid cysts. The Type II organisms utilize the serine pathway for carbon assimilation and have a complete TCA cycle. Facultative methane oxidizing bacteria have also been reported[21,22] which utilize not only methane but also some organic compounds. Biological and biochemical aspects of microbial growth on C_1 compounds were reviewed recently.[23] Fungal growth on C_1 compounds was reported with *Trichoderma lignorum*,[24] and also with *Streptomyces* sp.[25] Two C_1 compound assimilation pathways were observed in *Streptomyces* sp. The serine pathway operates at the early stage of growth phase. During the active logarithmic phase, carbon flows into cellular constituents from methanol via the pentose monophosphate pathway.

Microbial transformations of methane into methanol, formaldehyde, formate, and carbon dioxide are carried out by methane mono-oxygenase, methanol dehydrogenase (bacteria), or methanol oxidase (yeast), formaldehyde dehydrogenase, and formate dehydrogenase, respectively.

There have been several attempts in recent years to elucidate the mechanism of initial attack on methane by obligate methylotrophs. Initial indirect evidence for oxygenase involvement in the oxidation of methane came from ^{18}O-incorporation experiments.[11] Higgins and Quayle[26] succeeded in accumulating methanol during methane oxidation by methylotrophic bacteria in the presence of high concentrations of phosphate buffer. They demonstrated with ^{18}O that the oxygen in methanol was derived exclusively from molecular oxygen. Subsequently, there were two reports on NADH- and oxygen-dependent methane oxidation by cell-free particulate fractions from *Methylococcus capsulatus*, Texas strain[27] and *Pseudomonas methanica*.[28]

Tonge et al.[29] purified a three component methane mono-oxygenase enzyme system from *Methylosinus trichosporium* OB3b. The components are a soluble CO-binding cytochrome *c*, a copper-containing protein, and a low molecular weight protein. The methane mono-oxygenase from *M. trichosporium* OB3b also catalyzed the oxidation of ethane, *n*-propane, *n*-butane, and carbon monoxide. Contrary to that report, *Methylococcus capsulatus*, Bath strain contained a soluble methane-oxygenase system

(Colby et al.).[30] This soluble mono-oxygenase system was also resolved into three components:[31-33] Component A, molecular weight 200,000 with 2 g-atom Fe; Component B, a colorless protein with molecular weight 15,000; and Component C, molecular weight 44,600 with 1 FAD, 2 non-heme Fe, and 2 acid-labile S. However, in recent reports, Hou et al.[34] and Patel et al.[35] detected methane mono-oxygenase activity only in the cell-free particulate fraction, and not in the soluble fractions from various methane-utilizing bacteria. Thus, the location of methane mono-oxygenase in vivo remains to be clarified.

The yield of methanol from methane was approximately 4 μmol/hr/mg cells in the whole cell system and 8 μmol/hr/mg protein in the cell-free system. Methanol is produced industrially by the catalytic reaction of carbon monoxide and hydrogen which are made by reforming natural gas or from petroleum-derived heavy hydrocarbon mixture.

Studies of the catalytic function of metals at the active site(s) of methane mono-oxygenase may lead to a new generation catalyst which would catalyze the production of methanol from methane at higher efficiency and under milder reaction conditions.

Formaldehyde is produced from methanol by methanol dehydrogenase (EC 1.1.99.8) (bacteria) or methanol oxidase (EC 1.1.3.13) (yeast). The reactions are

in bacteria (in vitro)

$$RCH_2OH + PMS \rightarrow RCHO + PMSH_2$$

(where, R is hydrogen or an alkyl group and PMS is an artificial electron donor, phenazine methosulfate)

in yeast

$$CH_3OH + O_2 \rightarrow HCHO + H_2O_2$$

Methanol dehydrogenases have been studied extensively from: *Pseudomonas* M 27,[36] *Methylococcus capsulatus* Texas,[37] *Hyphomicrobium* W. C.,[38] *Pseudomonas* 2491,[39] *Hyphomicrobium* X,[40] *Pseudomonas* RJ1,[41] *Rhodopseudomonas acidophila,*[42] *Methylobacterium organophilum,*[43] *Pseudomonas* C,[44] *Methylomonas methanica,*[45] and *Methylosinus sporium.*[46]

Methanol oxidase was first purified by Ogata and his co-workers from *Kloeckera* sp.[47,48] and was shown to contain FAD. Subsequently, a similar enzyme was isolated from *Candida*[49] and *Hansenula.*[50]

Although the formation of formaldehyde can be demonstrated in vitro, the amount of formaldehyde accumulated in vivo is negligible, presumably due to its further oxidation.

Formic acid is produced from formaldehyde by formaldehyde dehydrogenase or from methane through the intermediate steps of methanol and formaldehyde. There are two types of formaldehyde dehydrogenases found in bacteria: (1) $NAD(P)^+$-dependent enzymes[51,52] and (2) $NAD(P)^+$-independent enzymes.[51,53-55] NAD^+-dependent formaldehyde dehydrogenase was also found in *Candida*[56,57] and *Kloeckera* sp.[58]

2. Other Gaseous Hydrocarbons

Ethane-utilizing bacteria were first reported by Bokova and his co-workers in 1947. Since then, several papers have been published.[9,12,59,60] Ethane is oxidized through ethanol, acetaldehyde, and acetic acid. Cell suspensions of the obligate methylotroph *Methylosinus trichosporium* OB3b were reported to oxidize ethane to acetone.[61] The intermediates in this reaction are acetoacetic acid and β-hydroxybutyric acid. Recently, Hou and Patel (unpublished results) identified many species, including newly isolated cultures and several known cultures, which can grow on ethane, propane, or *n*-butane as the sole sources for carbon and energy.

Propane- or butane-grown cells of *Mycobacterium smegmatis* transformed propane or *n*-butane to acetone or 2-butanone, respectively, via subterminal oxidation.[62] Pabst and Brown[63] presented evidence that isopropanol precedes both *n*-propanol and acetone in the catabolism of propane. Vestal and Perry[64] ruled out terminal oxidation of propane to propionic acid as an important step in the degradation of propane based on evidence that propane and propionic acid were metabolized by different pathways. The following pathway was then suggested for propane metabolism:

$$CH_3CH_2CH_3 \rightarrow CH_3\overset{\overset{\displaystyle OH}{|}}{C}HCH_3 \rightarrow CH_3\overset{\overset{\displaystyle O}{\|}}{C}CH_3 \rightarrow$$

$$CH_3\underset{\underset{\displaystyle O}{\|}}{C}CH_2OH \rightarrow CO_2 + CH_3COOH$$

However, further studies suggest that some terminal oxidation of propane does occur in *M. vaccae*.[65] Tanaka et al.[66] isolated a Gram-positive organism described as *Brevibacterium ketoglutamicum* that utilized C_1-C_4 gaseous alkanes. Blevins and Perry[67] reported that a Gram-positive rod, strain R-22, utilizes propane via methyl ketone and utilizes propionate via methyl malonyl-succinate. Propane-utilizing fungi, *Cunninghamella elegans* and *Penicillium zonatum* were also reported.[68]

Recently, we found that acetone was accumulated when a resting cell suspension of propane-grown cultures was incubated with propane (unpublished data). The cells also convert propane to both 1-propanol and 2-propanol. In further studies, it was found that resting-cell suspensions of *Norcardia paraffinica* ATCC 21198 or *Brevibacterium* sp. CRL 56 oxidized 1-propanol more rapidly than 2-propanol.

Butane is oxidized by *Arthrobacter, Brevibacterium,* and several fungi.[69,70] The oxidation is at the terminal position by *Mycobacterium vaccae*[71] to both butyrate and β-hydroxy butyrate.

3. Co-Oxidation and Stationary Transformation

Foster[72] discovered that *Pseudomonas methanica* oxidized ethane, propane, and butane while the organism metabolized its growth substrate, methane, and termed it as "co-oxidation". Over the years, the concept of co-oxidation has become somewhat confusing. The confusion is due to the extension by more than one author of the original definition of co-metabolism to cover the metabolism of nongrowth substrates in the absence of a true growth substrate. Perry[73] gave a thorough discussion on hydrocarbon co-oxidation. *M. vaccae* was reported to co-oxidize many cycloparaffins to the homologous cycloalkanones during propane oxidation.[74-76] Quayle[77] argued for a return to the original terminology and adoption of the term "stationary metabolism" to describe metabolism of nongrowth substrates in the absence of growth substrates. Use of the organism as a chemical cutting tool powered by an independent energy supply is a promising area of biotechnology.

In continuation of their epoxidation studies[78] Hou et al. at the Exxon Research Center discovered early in 1977 that resting-cell suspensions of methane-grown methylotrophs oxidized gaseous alkenes to their corresponding 1,2-epoxides. They also isolated many new methane-utilizing cultures which epoxidized 1-alkenes in the absence of methane.[34] The highest activity was observed on propylene and ethylene (10 μmol/hr/mg biomass). Liquid alkenes were not epoxidized. Cell-free preparations of these methane-grown cultures were also found to actively epoxidize 1-alkenes to their corresponding 1,2-epoxides.[35] The Exxon group also found that in the absence of methane, resting cell suspensions of methane-grown cultures oxidized alkanes to their corresponding primary alcohols, secondary alcohols, and methyl ketones. Since there is no co-substrate involved in these reactions, it seems more appropriate to name these "stationary transformations" than "co-oxidations". Both propylene oxide and 2-butanone are important high volume commodity chemicals. They are prepared currently by multi-step chemical processes which require high density energy.

Independently, Colby et al.[79] found that the crude form of methane mono-oxygenase they had isolated had a very wide substrate specificity and was capable of oxidizing alkanes, alkenes, alicyclic, heterocyclic, and aromatic compounds. Stirling and Dalton[80] have investigated the ability of whole cells to oxidize the 31 compounds which were oxidized by the cell-free methane monooxygenase preparations of *M. capsulatus* bath. Only five of these compounds were oxidized. They also observed considerable enhancement of the ability of whole cells to oxidize compounds in the presence of formaldehyde.

Contrary to the report of Stirling and Dalton,[80] Higgins[81] found that cell suspensions of *Methylosinus trichosporium* OB3b also oxidized a wide variety of substrates including alkanes, alkenes, ethers, and substituted methane derivatives. Recently, the Cetus Company announced a novel multi enzyme process for the oxidation of propylene.[82] The first enzyme reaction involves the conversion of olefin to halohydrin in the presence of halide, hydrogen peroxide, and an enzyme haloperoxidase. Haloperoxidase can be obtained from such as horseradish, seaweed, or *Caldariomyces*. In the second reaction, propylene halohydrin is transformed to propylene oxide by halohydrin epoxidase or by whole cells of *Flavobacterium* sp. "Cheap" hydrogen peroxide can be produced from a co-product process such as converting glucose to gluconic acid with glucose-1-oxidase. The following scheme shows overall reaction of cetus process.

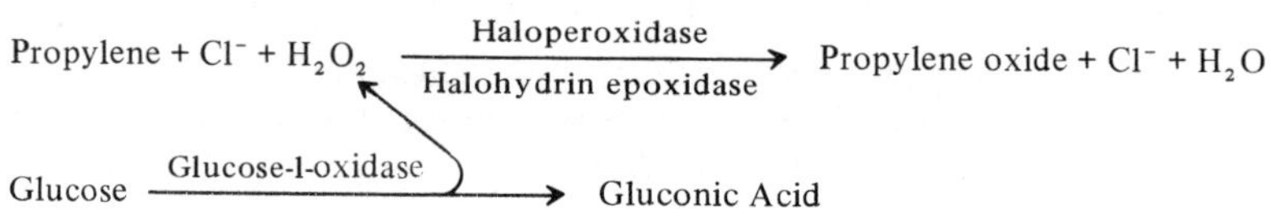

Another stationary transformation was demonstrated by Hou et al.[83] which showed that cell suspensions of methylotrophs as well as methanol-grown yeasts oxidized secondary alcohols to their corresponding methyl ketones. Among the secondary alcohols, 2-propanol and 2-butanol were oxidized at the highest rate, both at about 6 μmol/hr/mg cells. A novel secondary alcohol specific, alcohol dehydrogenase (SADH), was subsequently purified from both bacteria[84] and yeast.[85] SADH from both bacteria and yeasts are similar to each other in their molecular weight (95,000 daltons), subunits (2 subunits), metal content (2 g-atoms zinc), and cofactor requirement (NAD).

4. Liquid Alkanes

Liquid alkanes (C_5-C_{16}) have been utilized by selected strains of the genera *Pseudomonas, Arthrobacter, Acinetobacter,* yeasts, etc. Heptane,[86] dodecane,[87] tetradecane,[88-91] dodecane and tridecane,[92,95] and hexadecane[94] have been reported as growth substrates. Degradation of aliphatic hydrocarbons has recently been reviewed.[15]

Microbial transformations of liquid alkanes into detectable products have been reported. The initial attack on the liquid alkane molecules may be: (1) by terminal hydroxylation and subsequent oxidation to the corresponding carboxylic acid; (2) by subterminal hydroxylation in the ω-2 or ω-3 positions which is followed by the oxidation to the ketone. The intermediate primary and secondary alcohols may be esterified by acids and accumulate as end products in the medium.

Two mono-oxygenase systems which are responsible for the initial attack on the liquid alkane molecules have been studied in detail. The first was isolated from octane-grown *Pseudomonas oleovorans*[95-97] (renamed *Pseudomonas putida,*)[98] and consists of three protein components: rubredoxin, an NADH-rubredoxin reductase, and an ω-hydroxylase (Figure 1). The ability of *Ps. putida* to oxidize alkanes is encoded in a transmissible plasmid.[98-101] The possibility of a hydroperoxide ($R-CH_2OOH$) being produced as an intermediate in this reaction has been raised on several occasions.[102,103] However, this is thought to be unlikely, as hydroperoxides have never been isolated as intermediates.[15]

Another system involves cytochrome P450. It was first described for the hydroxylation of *n*-octane by *Corynebacterium* sp. 7EIC.[104,105] This system also consists of multiprotein components: a flavoprotein, an iron protein, a cytochrome P450, and an ω-hydroxylase (Figure 1).

Gallo et al.[106] found that in *Candida tropicalis* the alkane hydroxylase is specifically located in the microsomes and contains NADH-cytochrome c reductase in addition to the cytochrome P-450. Only *n* -alkanes with more than 14 carbon atoms are inducers of the synthesis of cytochrome P-450, the hydroxylase system,[107] the NADPH-cytochrome c reductase and cytochrome b_5.[108]

Decane was oxidized by *Corynebacterium* to 1-decanol, 1,10-decanediol, and decyl decanoate.[109] C_{12}-C_{16} *n* -alkanes were oxidized by both terminal and subterminal oxidation,[110] and were further converted to the corresponding ketones[111] or to esters of lower aliphatic acids.[112] Microbial transformations were proposed for use in the preparation of partial oxidation products of alkanes: nonane to nonanedioic acid,[113] decane to sebacic acid,[114] tetradecane to myristyl alcohols and its methyl ester;[115] penta- and hexadecane to the corresponding penta- and hexadecan-2, -3- and -4- ones;[116] hexadecane to tetradecanedioic acid.[119]

5. Subterminal Oxidation

Ladd[118] first suggested that subterminal oxidation may be an important oxidative route for microbes when he noted that an alkane-utilizing (C_8-C_{16}) *Corynebacterium* oxidized undecan-2-one faster than *n* -decylaldehyde. Klausmeier et al.[119] also postulated a nonterminal carbon oxidation by propane-utilizing mycobacterium. Leadbetter and Foster[11] found that *Pseudomonas methanica,* growing at the expense of methane, was able to co-oxidize propane and butane to their corresponding methyl ketones in addition to alcohols and acids. *Mycobacterium smegmatis* was able to grow on and produce the respective methyl ketones from propane, *n*-butane, *n*-pentane, or *n*-hexane.[62] Vestal and Perry[64] demonstrated subterminal oxidation of propane with *Brevibacterium* JOB5. Their conclusion was based on evidence indicating that propane and propionic acid were metabolized by different pathways. In higher alkanes, *n* -hexadecane was subterminally oxidized to ketones by *Arthrobacter* species.[120] Tridecane was oxidized to tridecan-2-ol and undecan-1-ol by *Pseudomonas aeruginosa,* thus suggesting initial attack at the C-2 atom.[121]

The subterminal oxidation of alkanes is not limited to bacteria. They are also demonstrated in *Penicillium* sp. and *Cunninghamella blakesleeana*[122] and in *Penicillium, Absidia glauca,* and *Mucor* sp.[110]

B. Alkenes

Different patterns of oxidation products indicate that the metabolic pathways for alkenes and *n* -alkanes are not interrelated. A variety of oxidation products from alk-1-enes have been isolated which suggests that the initial attack may be at either methyl ternimus or at the double bond,[123,124] The products isolated are, for example, 1,2-epoxide, 1,2-diol, ω-unsaturated fatty acid, ω-unsaturated primary and secondary alcohols, and 2-hydroxy acid; each of the same chain length as the alkene substrate.

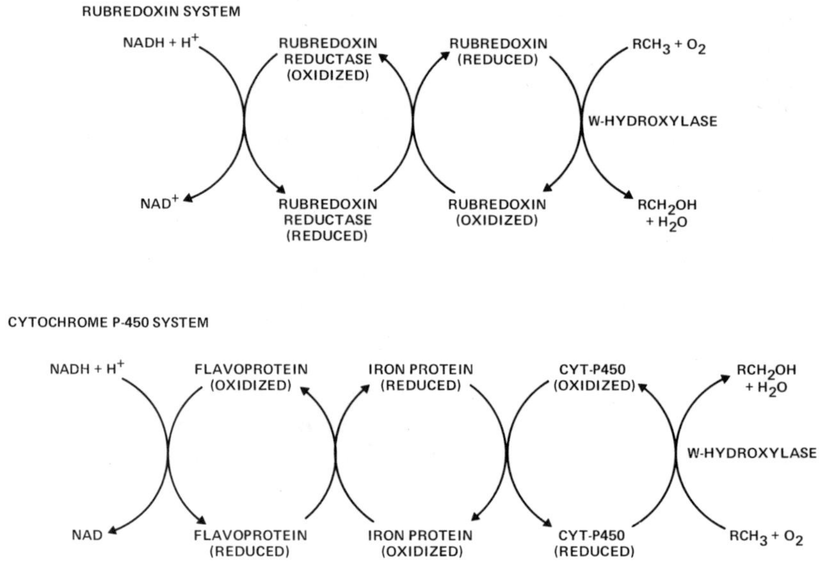

FIGURE 1. Microbial alkane mono-oxygenase systems.

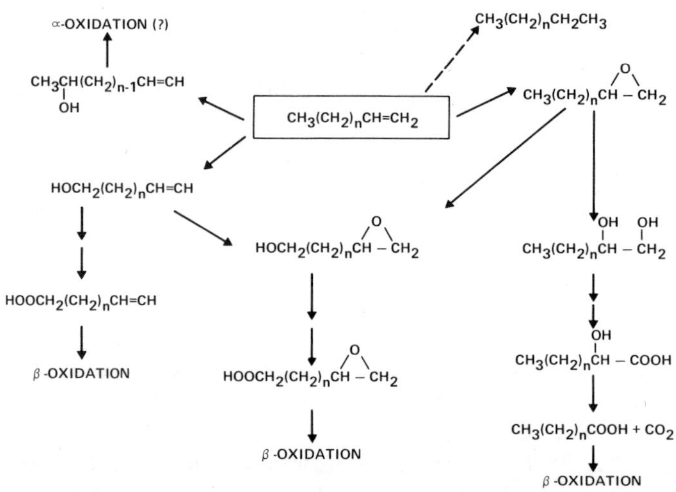

FIGURE 2. Microbial transformation of alkenes.

The epoxidation and the ω-hydroxylation are catalyzed by the same mono-oxygenase system.[125,126] The metabolic significance of the epoxide is not clear. Slow and spontaneous hydrolysis of the epoxide can occur to produce the corresponding diol.[13] In *Pseudomonas oleovorans,* there is apparently no further attack on the epoxide group and further degradation is via oxidation of the methyl group at the other end of the molecule.[78] In *Mycobacterium vaccae,*[127] the oxidation from both ends of the alk-1-ene molecule operates simultaneously. However, there was evidence of attack only at the methyl terminus of the alk-1-ene substrate by *Cunninghamella elegans*[128] and by *Corynebacterium simplex.*[129]

The gaseous olefin, propylene, is also oxidized diterminally. However, for successful growth on propylene, a C_2 + C_1 cleavage must occur across the double bond.[130] Pathways thought to be involved in the microbial transformation of alk-1-enes are shown in Figure 2.

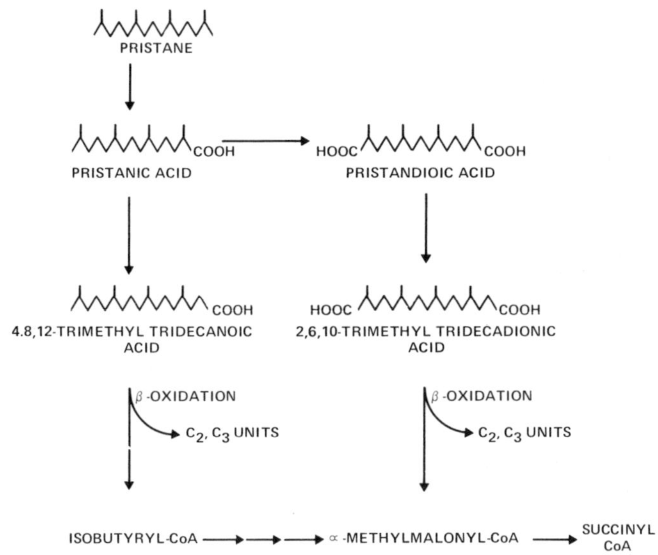

FIGURE 3. Microbial transformation of pristane. (From Pirnik, M. P., Atlas, R. M., and Bartha, R., *J. Bacteriol.*, 119, 875, 1974. With permission.)

Recently, DeBont[131] isolated a strain, *Mycobacterium* E20, from soil, which utilizes ethylene, ethylene oxide, or ethane as the sole source of carbon and energy for growth. Furthermore, whole cells of ethylene-grown *Mycobacterium* E20 oxidized not only ethylene but also ethylene oxide.[132] In addition, the whole cells could excrete ethylene oxide when ethylene was present.[62] The ethylene mono-oxygenase in the ethylene utilizing *Mycobacterium* E20 is different from the mono-oxygenase which catalyzes the oxidation of ethane.[133] It is also different from the nonspecific mono-oxygenase which catalyzes both the hydroxylation of alkanes and the epoxidation of alkenes described previously for methane- or octane-utilizing microbes.

Recently, Ohta and Tetsukawa[134] reported the asymmetric epoxidation of long chain terminal olefins by *Corynebacterium equi* IFO 3730. The product epoxide was confirmed to have 100% optical purity with (R) absolute configuration.

C. Branched-Chain Alkanes and Alkenes

It is known that only a few microorganisms utilize branched-chain hydrocarbons. The possible reason offered is that in most cases the oxidation enzymes are not capable of handling branched-chain substrates. Single branched alkanes are oxidized preferentially at the methyl group furthest from the branch point and the resulting fatty acids are then incorporated into the cell lipid.[124,127] Subsequent oxidation would be by the β-oxidation pathway and in some cases, by ω-oxidation.[135]

The oxidation of multiple branched alkanes such as pristane (2,3,10,14-tetramethyl pentadecane) has been studied by McKenna[136] using *Corynebacterium* sp. and later by Pirnik et al.[135] using *Brevibacterium erythrogenes.* The pathway is summarized in Figure 3. The initial ω-oxidation produces both pristanic and pristandioic acids. Subsequent β-oxidation yields propionyl-CoA and acetyl-CoA units alternately. The end products are either isobutyric-CoA or α-methyl malonyl-CoA. The former will then be converted to the latter, which will be isomerized to succinyl-CoA.

The oxidation of methyl branched alkenes such as squalene was studied by Yamada et al.[137,138] using *Arthrobacter* sp. Although one of the products, geranylacetone, was not further metabolized, a series of other products obtained indicates a progressive oxidation down towards intermediates of the tricarboxylic acid.

III. ALICYCLIC HYDROCARBONS

Alicyclic hydrocarbons are found in plant metabolites, the terpenoid hydrocarbons, and in crude oils, particularly in North Sea oil and in Athabasca tar oils.[139] Attempts to isolate a pure strain grown on pure alicyclic hydrocarbons have failed by many workers.[74,75,140] Instead, they found that these alicyclic hydrocarbons were co-oxidized by microorganisms growing on other substrates. An alkane-utilizing strain of *Mycobacterium vaccae* (JOB5) was shown by Ooyama and Foster[141] to co-oxidize a variety of alicyclic hydrocarbons to the corresponding ketones. A strain of *Pseudomonas aeruginosa* grown on *n* -heptane was also able to co-oxidize cycloalkanes to their corresponding alcohols.[140] More recently, a wide variety of strains of methane-grown microorganisms were found to be able to stationary-oxidize alicyclic hydrocarbons.[79,81] Microorganisms capable of growth on alicyclic alcohols and ketones are common in nature.[16] Therefore, cycloalkanes can be metabolized readily by mixed cultures of *n* - alkane utilizers which stationary-oxidize cycloalkanes and other microorganisms which are able to utilize the alicyclic oxidation products.

However, considerable progress has been made recently in the study of cyclohexane metabolism by pure bacterial strains. The growth of *Nocardia petroleophila* on methylcyclohexane[142] and of *Nocardia* sp. on either pure cyclohexane or on methylcyclohexane[143] were reported. An interesting feature of the *Nocardia* sp. was that when initially isolated, growth only occurred in conjunction with a strain of *Pseudomonas* that was incapable of growth on cyclohexane. Subsequently, this dependence upon the *Pseudomonas* was determined to be a source of biotin, as *Nocardia* is auxotrophic for biotin. Cyclohexane is oxidized to cyclohexanone via cyclohexanol. Further metabolism of the cyclohexanone is summarized in Figure 4. This pathway is common to several species.[143,145] Enzymes catalyzing the oxidation of cyclohexanone to 1-oxa-2-oxocycloheptane have been purified and studied in detail.[146,147]

n-Alkyl-substituted alicyclic hydrocarbons are more susceptible to microbial attack than the unsubstituted parent compounds. Methylcyclohexane and methylcyclopentane are present in crude oil in a significant amount and also accumulate in the environment.[148] Microbial degradation of these compounds is important in connection with waste treatment and pollution control. *n* -Octadecane grown *Nocardia* sp. was found to co-oxidize *n* -butylcyclohexane to cyclohexane acetic acid.[149] Beam and Perry[75] found that cyclohexane carboxylic acid and cyclohexane propionic acid were more suitable to microbial growth than cyclohexane acetic acid and cyclohexane butyric acid. Effective cleavage of an alicyclic ring only occurred when the side chain contained an odd number of carbon atoms.

Rho and Evans[150] isolated a strain of *Acinetobacter anitratum* that utilized cyclohexane carboxylic acid through 2-oxocyclohexane carboxylic acid. Some anaerobic photosynthetic strains,[151,152] as well as aerobic nonphotosynthetic strains,[16] also metabolize cyclohexane carboxylic acid by this pathway. From studies with a strain of *Corynebacterium cyclohexanicum,* Kaneda[153] postulated that cyclohexane carboxylate is metabolized through an aromatic intermediate (Figure 5). The aromatization of cyclohexane carboxylate was also reported by Blakley[154] and by Smith and Calley.[155] Aromatic compounds thus formed are further metabolized through either extra- or intradiol cleavage via the protocatechuate pathway. Since further metabolism of *p* -hydroxy phenylacetate is well established, it has been suggested that cyclohexane acetic acid, which is derived from C-even alkyl side chain cyclohehexane, is also metabolized through an analogous aromatization pathway.[16]

Major contributions to microbial transformation and metabolism of terpenoid hydrocarbons have been made by Bhattachrayya and his group. Their studies have attracted considerable attention from the flavoring and perfume industries. Microbial

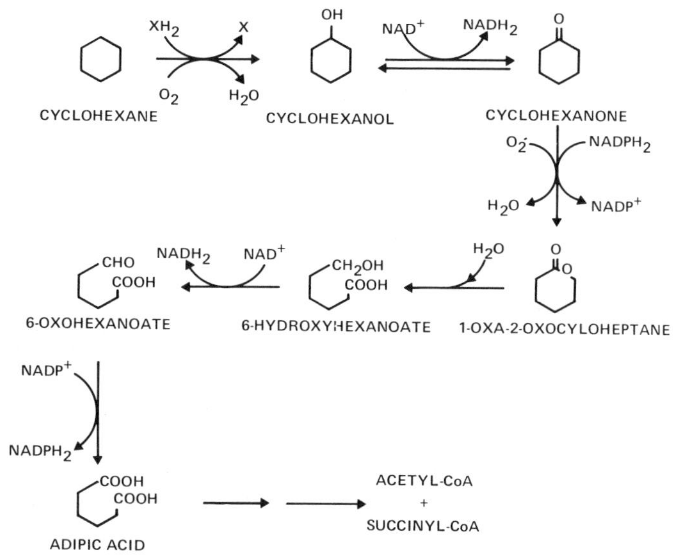

FIGURE 4. Microbial oxidation of cyclohexane.

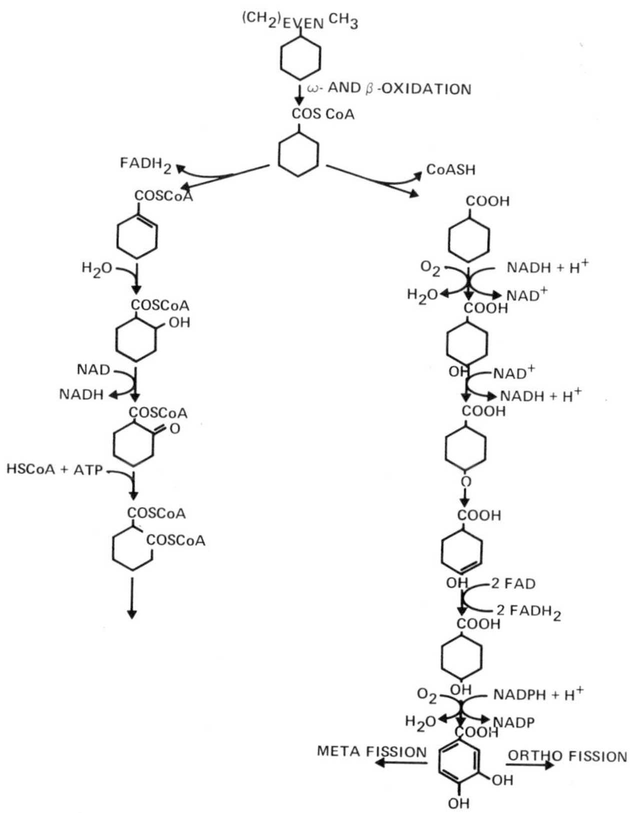

FIGURE 5. Microbial oxidation of C_{odd}-alkyl cyclohexane.

transformation of α- and β-pinenes by *Aspergillus niger*[156] and by *Pseudomonas* PL[157,158] were reported. Many intermediates were isolated. It seems that the initial microbial attack on pinenes occurs on both the ring and the substituted methyl group. Microbial transformation of limonene was also reported.[159,160] Other terpenoid hydrocarbons for which transformation pathways have been elucidated are camphene;[161-163] α-santalene;[164] menthene,[165-168] and menthane.[169]

IV. AROMATIC HYDROCARBONS

A. Mononuclear Aromatic Hydrocarbons

Coal and crude oil provide the main sources for aromatic hydrocarbons. Mononuclear aromatic hydrocarbons are commodity chemicals and are widely used in industry as, for example, solvents. World production figures for 1978, in millions of tons, are 11.4 for benzene, 6.1 for toluene, and 29.6 for xylenes. A large portion of these chemicals are partially oxidized nonbiologically to various chemical intermediates. Interest continues to grow in replacing these chemical oxidation processes with mild biotransformations or enzyme-mimicking processes. In addition, interest in the metabolism of these aromatic hydrocarbons has been stimulated by the concern over their effects on the environment.

1. Benzene

With *Pseudomonas rhodochrous* and *P. aeruginosa,* Marr and Stone[170] first demonstrated that benzene is metabolized through catechol and *cis, cis*-muconic acid. Gibson and his colleagues subsequently found that *cis*-benzene glycol is an intermediate between benzene and catechol.[171,172] Resting cells of a glucose-grown, benzene-induced mutant strain of *Pseudomonas putida* 39/D produced up to 1.1 mg/mℓ of *cis*-benzene glycol in a medium containing benzene.[173] This is another example of "stationary transformation" with industrial significance that grows cells on an inexpensive substrate while conserving the cosubstrate primarily for product formation. The main differences between bacterial and mammalian systems for benzene metabolism is that, in the latter system, benzene is oxidized to benzene 1,2-oxides, and then to *trans* - benzene glycol before forming catechol.

The bacterial enzyme systems involved in the oxidation of benzene are extremely labile.[170,171,174] They incorporate dioxygen into the benzene molecule. A stable benzene-oxidizing enzyme system from *Pseudomonas putida* has been purified and characterized recently.[175,176] This system consists of three proteins: an FAD-containing protein with a molecular weight of about 60,000; a small protein of 21,000 containing nonheme iron; and a red-colored protein of 186,000 containing nonheme iron. The electron donor of this system is NADH. The bacterial transformation of benzene to catechol and the mechanism of benzene dioxygenase are summarized in Figure 6.

2. Substituted Benzene

Microbial oxidation of toluene, the simplest hydrocarbon derivative of benzene, is initiated through either the methyl group or the aromatic ring, depending upon the organism employed. With *Pseudomonas aeruginosa,* Kitagawa[177] first reported the oxidation of toluene at the methyl group via benzylalcohol, benzaldehyde, benzoic acid, and catechol intermediates (Figure 7). This pathway was supported by the results of an investigation by Nozaka and Kusunose.[178] Extracts of *P. aeruginosa* grown on xylene could also oxidize toluene. ^{14}C-benzoic acid was obtained with ^{14}C-toluene as the substrate. The same pathway was also found in *P. putida (arvilla)* mt-2 for the oxidation of toluene and xylene (Figure 7).[178] The resulting benzoic acid or toluic acids are further metabolized via catechol and methyl-substituted catechol intermediates, respectively.[179]

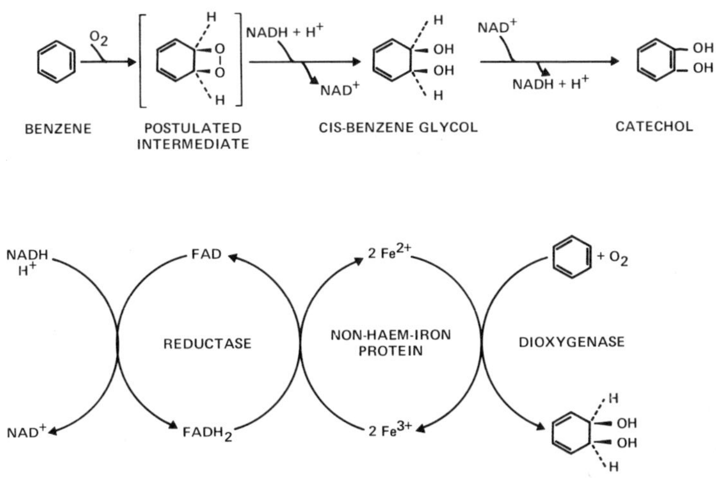

FIGURE 6. Microbial transformation of benzene to catechol and the mechanism of benzene dioxygenase.

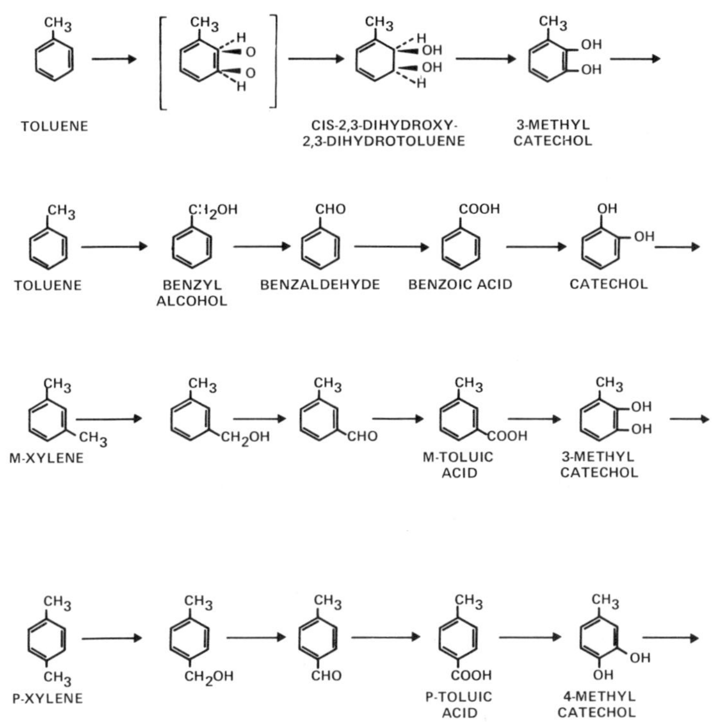

FIGURE 7. Pathways for toluene and xylene biotransformation.

The oxidation of toluene through the aromatic ring, analogous to those involved in benzene oxidation yielding 3-methyl catechol, was reported by Claus and Walker[180] using other species of *Pseudomonas* and *Achromobacter* (Figure 7). The fact that 20 other toluene- or cresol-oxidizing microbes all oxidized 3-methyl catechol immediately and benzoic acid only after a lag, suggests that this is a widely used pathway for toluene metabolism. Additional evidence for this pathway has been reported by Nozaka and Kusunose[181] with *Pseudomonas mildenberger* and by Gibson et al.[182] with *Pseudo-*

monas putida. Experiments with cell extracts showed that, similar to the benzene dioxygenases, the toluene-oxidizing activity can be separated into three protein fractions. The activity requires Fe^{2+} together with NADH.[183] Toluene dioxygenase from *Pseudomonas putida* was purified by affinity chromatography and was identified as an iron-sulfur protein.[184]

The oxidation of *m* - and *p* -xylenes at the methyl group to yield 3-, and 4-methyl catechol, respectively, was reported by Omori et al.[185-188] and by Davey and Gibson.[189] Direct oxidation of the aromatic ring of *p* - and *m* -xylenes by *P. putida* 39/D through their *cis*-dihydrodiol intermediates was also reported.[190] However, it has not been possible to isolate microorganisms that grow on *o* -xylene.[187,191,192] The oxidation of *0* -xylene to *o* -toluene was demonstrated in a *Nocardia* sp. only with the co-oxidation technique.[193]

The oxidation of other aromatic hydrocarbons by microorganisms is generally found to be similar to the types of reaction described for benzene, toluene, and xylene metabolism. Thus, ethylbenzene is oxidized through the aromatic ring attack to 3-ethylcat echol by *P. putida*.[171] A similar pathway was suggested by Baggi et al.[194] for 2-phenyl butane, 3-phenylpentane, and 4-phenylheptane metabolism in *P. putida* and *P. acidovorans*. Other examples for the aromatic ring attack are isopropylbenzene, isobutylbenzene, and *n* -butylbenzene by *P. desmolytica* (S449B1) and *P. convexa* (S107B1).[191,192,195] Production of *β*-phenethyl alcohol from styrene was reported by Sielicki et al.[196] and by Shirai and Hisatsuka.[197] Recently, Jigami et al.[198] showed the coexistence of different pathways in the metabolism of *n* -propylbenzene by *Pseudomonas* sp. *n* -Propylbenzene was oxidized to *β*-phenyl propionic acid and benzoic acid by initial oxidation of the *n* -propyl side chain. The same strain also oxidized *n* -propylbenzene to 3-*n* -propyl catechol by initial oxidation of positions 2 and 3 of the aromatic nucleus.

Long-chain alkylbenzenes are oxidized at the terminal methyl group to the corresponding phenyl alkanoic acids. The growth of *Nocardia* species on *n* -decylbenzene, *n* -dodecylbenzene, and *n* -octadecylbenzene,[199] and *n* -nonylbenzene and *n* -dodecylbenzene[136,149] was demonstrated. As the alkyl chain length grows and the substituent becomes the major part of the molecule, these compounds are more realistically regarded as substituted alkanes rather than substituted benzenes. Thus, like the *n* -alkanes, they undergo oxidation to give phenyl alkanoic acids that are then metabolized by oxidation. In some cases, this breakdown of the alkyl group alone is sufficient to support growth and the organism may not always be able to degrade the benzene substituent, which can be accumulated as, for example, phenylacetic acid,[199] phenylpropionic, or cinnamic acid.[200] *n* -Alkylbenzenes are oxidized by yeasts in a similar way.[201]

A great deal is known about hydroxylases and particularly about the enzyme from *P. oleovorans*, which consists of three components (Figure 1). However, there is little information on the nature of oxygenases responsible for initiating the oxidation of the methyl group in aromatic substrates. Nozaka and Kusunose[178] resolved the *p* -xylene hydroxylase into two fractions by DEAE-cellulose chromatography. They found that this xylene hydroxylase system was unable to attack decane or hexadecane. The hydroxylase of *Ps. aeruginosa* could not oxidize toluene or xylene.[202] The hydroxylase from *Corynebacterium* 7EIC grown on octane was also found to bc inactive on substituted benzenes, toluene, and *m* - and *p* -xylenes.[203] On the contrary, similarities between these two hydroxylase systems, the oxidation of *n* -alkane, and the oxidation of the terminal methyl group of alkylbenzene, were reported. The *n* -alkane hydroxylase system from *Ps. aeruginosa* also oxidized toluene to benzyl alcohol.[204] The ring dioxygenase of Gibson et al.[205] was also able to convert ethylbenzene to phenyl ethanol.

Although all of the hydrocarbon hydroxylases reported are oxygenases, Hopper[206] recently discovered a methyl-group hydroxylase which converts *p* -cresol to *p* -hydroxylbenzylalcohol anaerobically.

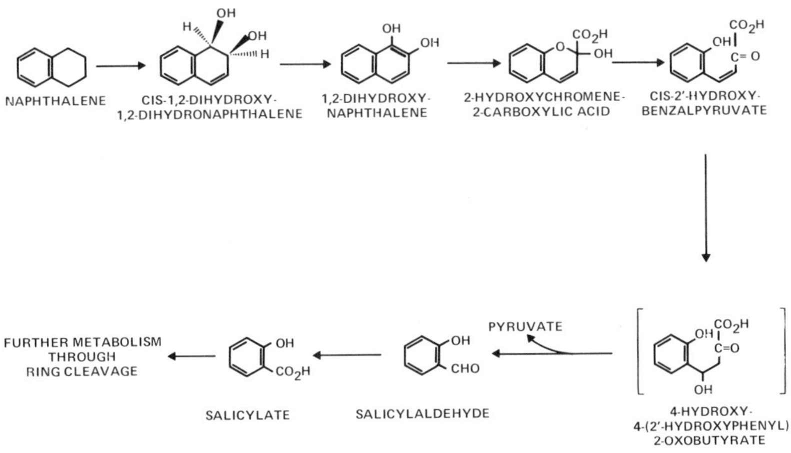

FIGURE 8. The metabolic pathway of naphthalene in Pseudomonads.

3. Aromatic Ring-Fission

Catechol or substituted catechols are the products of initial aromatic hydrocarbon transformation. Ring-fission of these compounds occurs by three distinct pathways, depending upon the substrates and microbes.[207] The intradiol pathway involves cleavage of the ring between the two hydroxy groups. There are two types of extradiol ring cleavage: proximal and distal extradiol cleavages. These pathways have been reviewed by Dagley[208] and by Chapman.[209]

B. Polynuclear Aromatic Hydrocarbons

Polynuclear aromatic hydrocarbons, as well as their derivatives, exist in natural sources, particularly in coal, crude oil, and plant materials. Examples of unsaturated polyaromatic hydrocarbons are biphenyl, naphthalene, anthracene, phenanthrene, benzo[a]anthracene, crysene, benzo[a]pyrene, and dibenzo[a,h]anthracene. Sulfur- and nitrogen-containing polynuclear aromatic hydrocarbons are also common in nature. The presence of sulfur and nitrogen result in catalyst deactivation, which is a big problem for the petroleum industry. Furthermore, some of these hydrocarbons and their metabolites have been found to be potential carcinogens.

In contrast to the liver microsomal system and some eukaryotic microorganisms,[210] procaryotic microorganisms oxdize polynuclear aromatic hydrocarbons through dioxetane and *cis*-diol intermediates analogous to those occurring with benzene. The initial oxygenative attack on polynuclear aromatic hydrocarbons by microorganisms also involves a dioxygenase enzyme.

Naphthalene, the simplest polyaromatic hydrocarbon, is produced domestically either from petroleum or coal raw materials and is available in multimillion pound quantities (600 million lb in 1973).[213] Early work on microbial oxidation of naphthalene was conducted by Davis and Evans.[212] Based on their findings, a pathway of naphthalene metabolism was proposed (Figure 8). Recently, Barnsley[213] found that 1,2-dihydroxynaphthalene is oxidized through 2-hydroxychromene-2-carboxylic acid to *cis* -2-hydroxybenzal-pyruvate. This compound is then oxidized through salicylaldehyde, salicylate, and then catechol. Catechol can undergo either ortho or meta ring-fission depending on the organism.

Industrial research on biological production of products from naphthalene has concentrated on the production of salicylic acid. Some fungal species metabolize aromatic hydrocarbons to metabolites identical to those formed by mammalian enzymes.[214,215]

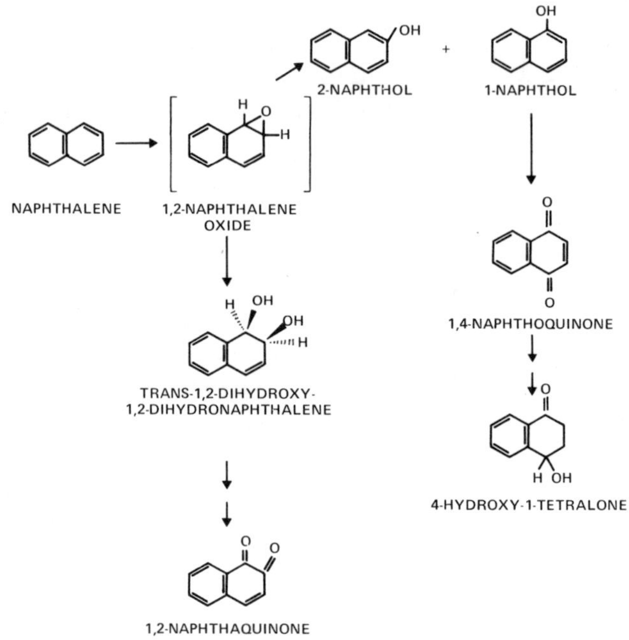

FIGURE 9. The metabolic pathway of naphthalene by Cunninghamella. (From Cerniglia, C. E. and Gibson, D. T., *Arch. Biochem. Biophys.*, 34, 363, 1977. With permission.)

Wegner[216] described the biotransformation of naphthalene and of substituted naphthalenes to their corresponding diols with *Nocardia* sp. Recently, Cerniglia and Gibson[217,218] reported that *Cunninghamella elegans* oxidizes naphthalene through 1-naphthol, 2-napththol, and trans-1,2-dihydroxy-1, 2-dihydronaphthalenes (Figure 9). Naphthalene metabolism was found in many genera from the major fungal taxa.[217] Algal oxidation of naphthalene was reported to be through 1-naphthol and *cis*-1,2-hydroxy-1, 2-dihydronaphthalene intermediates.[219-221]

Alkyl-substituted naphthalenes are oxidized by pathways analogous to those for the unsubstituted parent hydrocarbons. Thus, 1-methyl- and 2-methyl-naphthalene are oxidized to 3-methyl- and 4-methylsalicylic acid, respectively. However a recent report indicates that alkyl-group oxidation also occurs.[222]

Other examples of aromatic ring attack were reported on dibenzothiophene with *Pseudomonas*.[223,224]

The microbial oxidation of phenanthrene and anthracene was studied by Evans and his collaborators.[225] The metabolic pathways of these trinuclear aromatic hydrocarbons (Figure 10 for phenanthrene, and Figure 11 for anthracene) show many similarities to the sequence described for naphthalene. Both 1,2-dihydroxynaphthalene and 2,3-dihydroxynaphthalene are further metabolized through the naphthalene pathway. Recently, Kiyohara et al.[226] described an alternative pathway of phenanthrene metabolism. *Aeromonas* sp. metabolizes 1-hydroxy-2-naphthoic acid, an intermediate of phenanthrene metabolism, through *o*-phthalic acid and protocatechuic acid but not through salicylate or catechol. *Beijerinckia* sp. also oxidizes phenanthrene and anthracene.[227]

Benzo[*a*]pyrene and benzo[*a*]anthrancene are also oxidized by *Beijerinckia* sp. into polar products with *cis*-configuration.[228] These observations lead to a suggestion that *cis*-hydroxylation of aromatic hydrocarbons is a common reaction in prokaryotic microorganisms.[229] However, fungi oxidize these compounds by a mechanism similar to

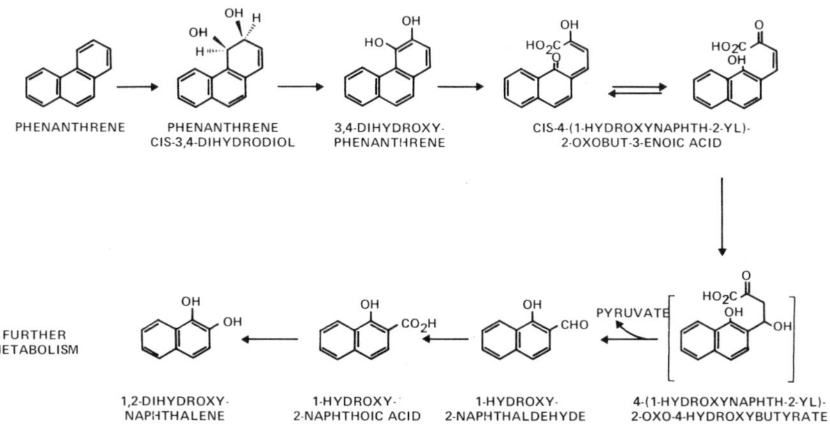

FIGURE 10. Microbial transformation of phenanthrene.

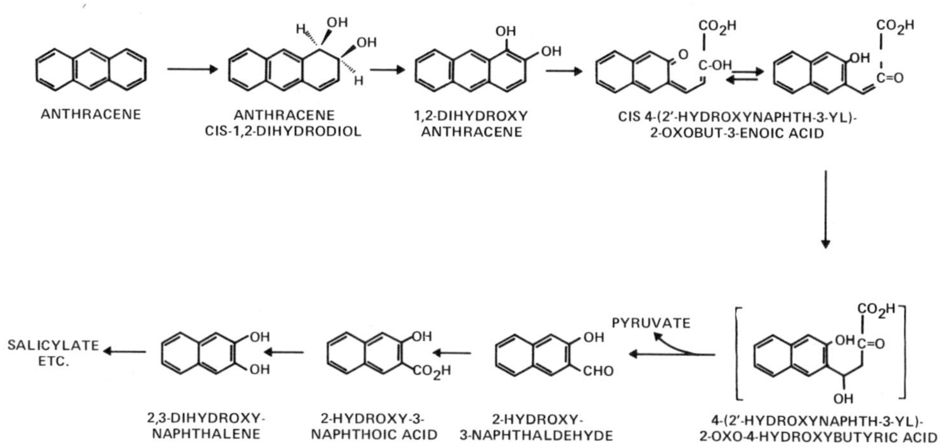

FIGURE 11. Microbial transformation of anthracene.

that observed in mammalian systems. Benzo[*a*]pyrene is oxidized by *Cunninghamella elegans* to polar products with trans-configuration.[230] Recently, production of a red benzo[*a*]anthracene from kerosene by *Pseudomonas* sp. S7K5 was reported.[231]

In bacteria, it has been shown that a dioxygenase enzyme initiates the oxidative attack on biphenyl to form 2,3-dihydroxybiphenyl. Further oxidation of 2,3-dihydroxybiphenyl can occur by two routes, depending upon the organism (Figure 12).[232-235] There is some evidence that fungi metabolize biphenyl to metabolites similar to those formed by mammalian systems. Biphenyl is oxidized to 2-hydroxy-, 3-hydroxy-, or 4-hydroxy-biphenyl.[215-236]

V. FUTURE PROSPECTS

Although progress in microbial hydrocarbon technology has been relatively slow due to the low solubility of hydrocarbons in water, the study of hydrocarbon biotransformations has been highly rewarding scientifically. It has uncovered, for example, the biochemistry of oxygenase enzymes, aerobic and anaerobic ring cleavage mechanisms, and the genetics and regulation of aliphatic and aromatic hydrocarbon transformation. From the application point of view, the hydrocarbon-utilizing microorganisms may some day be used in place of the organic chemist to modify pre-existing hydrocarbons or to remove unwanted hydrocarbons.

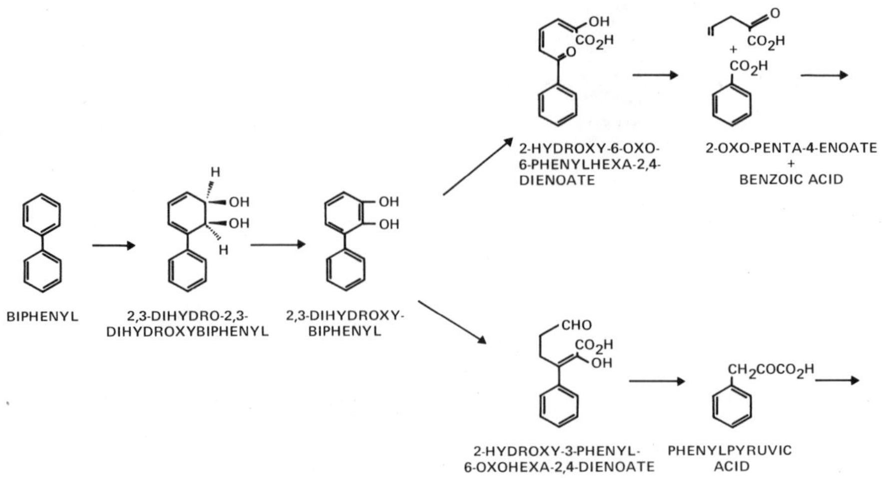

FIGURE 12. Microbial transformation of biphenyl.

Aside from the traditional fermentation and growth of microbes on hydrocarbons, co-oxidation and stationary transformation techniques and genetic engineering are most promising areas of hydrocarbon biotechnology for the future.

Co-oxidation and stationary transformation offer valuable methods for obtaining new products and for increasing yields of other products. They also represent some ecological systems operating in nature. From an industrial standpoint, it is ideal to use an organism (or its enzymes) as a chemical cutting tool powered by an independent (renewable) energy supply. Application of this concept has been emphasized on the partial oxidation of aromatic hydrocarbons by species of *Nocardia* during growth on *n* -alkanes.[237] Other examples of co-oxidation were reviewed by Abbott and Gledhill[2] and by Gibson.[238] Recently, methane-grown microorganisms were found to oxidize gaseous alkenes into their epoxides and gaseous alkanes into their corresponding alcohols and methyl ketones.[34,79,81,83] Other breakthroughs in this area are expected in the near future.

In 1972, Chakrabarty[239] showed that the genes coding for the transformation of certain hydrocarbons can be carried on extrachromosomal DNA (plasmids). These plasmids can be transferred between bacteria, both in vivo and in vitro. The hydrocarbon degradative plasmids demonstrated in *Pseudomonas putida* thus far are: camphor,[240] octane and other straight-chain alkanes,[241-244] naphthalene,[245] and toluene, *m*- and *p* -xylenes.[179,246,247] It is possible to incorporate several phenotypic characters into one bacterium. Friello et al.[248] have transferred three of the hydrocarbon plasmids (NAH, TOL, and CAM-OCT) into one strain of *Pseudomonas aeruginosa* to give a multiplasmid *Pseudomonas* MPP. This strain is potentially able to transform simultaneously some linear alkanes, aromatic and polyaromatic hydrocarbons. This multihydrocarbon-degrading bacterium might be useful in the cleanup of oil spills or of wastes resulting from industrial pollution. Microbial genetics relating to hydrocarbon degradation has recently been reviewed by Williams.[249] The advances in gene splicing and genetic engineering in recent years have opened up new possibilities for constructing laboratory bacterial strains with new and advantageous characteristics capable of hydrocarbon transformation.

REFERENCES

1. **Kieslich, K.,** Microbial transformations of nonsteroid cyclic compounds, John Wiley & Sons, New York, 1975.
2. **Abbott, B. J. and Gledhill, W. E.,** The extracellular accumulation of metabolic products by hydrocarbon-degrading microorganisms, *Adv. Appl. Microbiol.,* 14, 249, 1971.
3. **Fonken, G. S. and Johnson, R. A.,** *Chemical Oxidations with Microorganisms,* Marcel Dekker, New York, 1972.
4. **Skryabin, G. K. and Golovleva, T. A. M.,** *Microorganisms in Organic Chemistry,* Nauka Press, Moscow, 1976.
5. **Beukers, R., Marx, A. F., and Zuidweg, M. H. J.,** Microbial conversion as a tool in the preparation of drugs, in *Drug Design,* Vol. 3, Ariens, E.J., Ed., Academic Press, New York, 1973, 131.
6. **Sebek, O. K. and Kieslich, K.,** Microbial transformations in Organic Compounds, Annual Report of Fermentation Process, 1977, 267.
7. **Sih, C. J. and Rosazza, J. P.,** Microbial transformations in organic synthesis, *Tech. Chem.,* 10, 69, 1976.
8. **Miyoshi, M.,** Die durehbohrung von membranen durch Pilzbaden, *Jahrb. Wiss Bot.,* 28, 269, 1895.
9. **Dworkin, M. and Foster, J. W.,** Experiments with some microorganisms which utilize ethane and hydrogen, *J. Bacteriol.,* 75, 592, 1958.
10. **Leadbetter, E. R. and Foster, J. W.,** Studies on some methane-utilizing bacteria, *Arch. Mikrobiol.,* 30, 91, 1958.
11. **Leadbetter, E. R. and Foster, J. W.,** Oxidation products formed from gaseous alkanes by the bacterium *Pseudomonas methanica, Arch. Biochem.,* 82, 491, 1959.
12. **Leadbetter, E. R. and Foster, J. W.,** Bacterial oxidation of gaseous alkanes, *Arch. Mikrobiol.,* 35, 92, 1960.
13. **Van der Linden, A. C. and Thijssee, G. J. E.,** The mechanisms of microbial oxidation of hydrocarbons, *Adv. Enzym.,* 27, 469, 1965.
14. **Markovetz, A. J.,** Subterminal oxidation of aliphatic hydrocarbons by microorganisms, *Crit. Rev. Microbiol.,* 1, 225, 1972.
15. **Ratledge, C.,** Degradation of aliphatic hydrocarbons, in *Developments in Biodegradation of Hydrocarbons - I,* Watkinson, R. J., Ed., Applied Science Publishers, 1978, 1.
16. **Trudgill, P. W.,** Microbial degradation of alicyclic hydrocarbons, in Developments in Biodegradation *of Hydrocarbons — I,* Watkinson, R. J., Ed., Applied Science Publishers, 1978, 47.
17. **Hopper, D. J.,** Microbial degradation of aromatic hydrocarbons, in *Developments in Biodegradation of Hydrocarbons — I,* Watkinson, R. J., Ed., Applied Science Publishers, 1978, 85.
18. **Watkinson, R. J. and Cripps, R. E.,** Polycyclic aromatic hydrocarbons: metabolism and environmental aspects, in *Developments in Biodegradation of Hydrocarbons — I,* Watkinson, R. J., Ed., Applied Science Publishers, 1978, 113.
19. **Higgins, I. J. and Gilbert, P. D.,** The biodegradation of hydrocarbons, in *The Oil Industry and Microbial Ecosystems,* Chater, K. W. A. and Somerville, H. J., Eds., Heyden and Son Ltd. Press, 1978, 80.
20. **Whittenbury, R., Phillips, K. C., and Wilkinson, J. F.,** Enrichment, isolation and some properties of methane-utilizing bacteria, *J. Gen. Microbiol.,* 61, 205, 1970.
21. **Patt, T. E., Cole, G. C., Bland, J., and Hanson, R. S.,** Isolation and characterization of bacteria that grow on methane and organic compounds as sole source of carbon and energy, *J. Bacteriol.,* 120, 955, 1974.
22. **Patel, R. N., Hou, C. T., and Felix, A.,** Microbial oxidation of methane and methanol: isolation of methane-utilizing bacteria and characterization of a facultative methane-utilizing isolate, *J. Bacteriol.,* 136, 352, 1978.
23. **Colby, J., Dalton, H., and Whittenbury, R.,** Biological and biological aspects of microbial growth on C_1 compounds, *Ann. Rev. Microbiol.,* 33, 481, 1979.
24. **Tye, R. and Willets, A.,** Fungal growth on C_1 compounds, *Ann. Rev. Microbiol.,* 33, 481, 1979.
24. **Tye, R. and Willets, A.,** Fungal growth on C_1 compounds: quantitative aspects of growth of a methanol-utilizing strain of *Trichoderma lignorum* in batch culture, *Appl. Environ. Microbiol.,* 33, 758, 1977.
25. **Kato, N., Tsuji, K., Ohashi, H., Tani, Y., and Ogata, K.,** Two assimilation pathways of C_1 compound in *Streptomyces* sp. No. 239 during growth on methanol, *Agr. Biol. Chem.,* 41, 29, 1977.
26. **Higgins, I. J. and Quayle, J. R.,** Oxygenation of methane by methane-grown *Pseudomonas methanica* and *Methanomonas methanooxidans, Biochem. J.,* 118, 201, 1970.
27. **Ribbons, D. W.,** Oxidation of C_1 compounds by particulate fractions from *Methylococcus capsulatus:* distribution and properties of methane-dependent reduced nicotinamide adenine dinucleotide oxidase (methane hydroxylase), *J. Bacteriol.,* 122, 1351, 1975.

28. **Ferenci, T., Strom, T., and Quayle, J. R.,** Oxidation of carbon monoxide and methane by *Pseudomonas methanica, J. Gen. Microbiol.,* 91, 79, 1975.
29. **Tonge, G. M., Harrison, D. E. F., and Higgins, I. J.,** Purification and properties of the methane monooxygenase enzyme system from *Methylosinus trichosporium* OB3b, *Biochem. J.,* 161, 333, 1977.
30. **Colby, J. and Dalton, H.,** Some properties of a soluble methane mono-oxygenase from *Methylococcus capsulatus* strain bath, *Biochem. J.,* 157, 495, 1976.
31. **Colby, J. and Dalton, H.,** Resolution of the methane mono-oxygenase of *Methylococcus capsulatus* (Bath) into three components. Purification and properties of component c, a flavoprotein, *Biochem. J.,* 171, 461, 1978.
32. **Stirling, D., Colby, J., and Dalton, H.,** A comparison of the substrate and electron-donor specifications of the methane monooxygenases from three strains of methane-oxidizing bacteria, *Biochem. J.,* 177, 361, 1979.
33. **Colby, J. and Dalton, H.,** Characterization of the second prosthetic group of the flavoenzyme NADH-acceptor reductase (component c) of the methane monooxygenase from *Methylococcus capsulatus* (Bath), *Biochem. J.,* 177, 903, 1979.
34. **Hou, C. T., Patel, R. N., Laskin, A. I., and Barnabe, N.,** Microbial oxidation of gaseous hydrocarbons: epoxidation of C_2 to C_4 n-alkenes by *Methylotrophic* bacteria, *Appl. Environ. Microbiol.,* 38, 127, 1979.
35. **Patel, R. N., Hou, C. T., Laskin, A. I., Felix, A., and Derelanko, P.,** Microbial oxidation of gaseous hydrocarbons. II. Hydroxylation of alkanes and epoxidation of alkenes by cell-free particulate fractions of methane-utilizing bacteria, *J. Bacteriol.,* 139, 675, 1979.
36. **Anthony, C. and Zatman, L. J.,** The microbial oxidation of methanol. Purification and properties of the alcohol dehydrogenase of *Pseudomonas* sp. M_{27}, *Biochem. J.,* 104, 953, 1967.
37. **Patel, R. N., Bose, H. R., Mandy, W. J., and Hoare, D. S.,** Physiological studies of methane and methanol-oxidizing bacteria: comparison of a primary alcohol dehydrogenase from *Methylococcus capsulatus* (Texas) and *Pseudomonas* sp. M_{27}, *J. Bacteriol.,* 110, 570, 1972.
38. **Sperl, G. T., Forrest, H. S., and Gibson, D. T.,** Substrate specificity of the purified primary alcohol dehydrogenase from methanol-oxidizing bacteria, *J. Bacteriol.,* 118, 541, 1974.
39. **Yamanaka, K. and Matsumoto, T.,** Purification, crystallization and properties of primary alcohol dehydrogenase from methanol-oxidizing *Pseudomonas* sp. No. 2941, *Agri. Biol. Chem.,* 41, 467, 1977.
40. **Duine, J. A., Frank, J., and Westerling, J.,** Purification and properties of methanol dehydrogenase from *Hyphomicrobium* X, *Biochim. Biophys. Acta,* 524, 277, 1978.
41. **Mehta, R. J.,** Studies on methanol oxidizing bacteria. II. Purification and properties of methanol dehydrogenase from *Pseudomonas* RJ1, *Antonie van Leeuwenhoek,* 39, 295, 1973.
42. **Bamforth, C. and Quayle, J. R.,** The dye-linked alcohol dehydrogenase of *Rhodopseudomonas acidophila.* Comparison with dye-linked methanol dehydrogenases, *Biochem. J.,* 169, 677, 1978.
43. **Wolf, H. L. and Hanson, R. S.,** Alcohol dehydrogenase from *Methylobacterium organophilum, Appl. Environ. Microbiol.,* 36, 105, 1978.
44. **Goldberg, J.,** Purification and properties of a methanol-oxidizing enzyme in *Pseudomonas* C, *Eur. J. Biochem.,* 63, 233, 1976.
45. **Patel, R. N., Hou, C. T., and Felix, A.,** Microbial oxidation of methane and methanol: crystallization of methanol dehydrogenase and properties of holo- and apo-methanol dehydrogenase from *Methylomonas methanica, J. Bacteriol.,* 128, 413, 1978.
46. **Patel, R. N. and Felix, A.,** Microbial oxidation of methane and methanol: crystallization and properties of methanol dehydrogenase from *Methylosinus sporium, J. Bacteriol.,* 128, 413, 1976.
47. **Tani, Y., Miya, T., Nishikawa, H., and Ogata, K.,** The microbial metabolism of methanol. I. Formation and crystallization of methanol-oxidizing enzyme in a methanol-utilizing yeast, *Kloeckera* sp. No. 2201, *Agr. Biol. Chem.,* 36, 68, 1972.
48. **Tani, Y., Miya, T., and Ogata, K.,** The microbial metabolism of methanol. II. Properties of crystalline alcohol oxidase from *Kloeckera* sp. No. 2201, *Agric. Biol. Chem.,* 36, 76, 1972.
49. **Sahm, H. and Wagner, F.,** Microbial assimilation of methanol: the ethanol- and methanol-oxidizing enzymes of the yeast *Candida boidinii, Eur. J. Biochem.,* 36, 250, 1973.
50. **Kato, N., Omori, Y., Yani, Y., and Ogata, K.,** Alcohol oxidases of *Kloeckera* sp. and *Hansenula polymorpha.* Catalytic properties and subunit structures, *Eur. J. Biochem.,* 64, 341, 1976.
51. **Boulton, C. A. and Large, P. J.,** Synthesis of certain assimilatory and dissimilatory enzymes during bacterial adaptation to growth on trimethyl amine, *J. Gen. Microbiol.,* 101, 151, 1977.
52. **O'Connor, M. L. and Hanson, R. S.,** Enzyme regulation in *Methylobacterium* organophilum, *J. Gen. Microbiol.,* 101, 327, 1977.
53. **Mehta, R. J.,** Studies on methanol-oxidizing bacteria. II. Purification and properties of methanol dehydrogenase from *Pseudomonas* RJ, *J. Microbiol. Serol.,* 41, 89, 1973.

54. **Patel, R. N., Hou, C. T., and Felix, A.,** Microbial oxidation of methane and methanol: and properties of a heme-containing aldehyde dehydrogenase from *Methylomonas methylovora, Arch. Microbiol.,* 122, 241, 1979.
55. **Patel, R. N., Hou, C. T., Felix, A., and Derelanko, P.,** Purification and properties of a heme-containing aldehyde dehydrogenase from *Methylosinus trichosporium, Arch. Biochem. Biophys.,* 1980, in press.
56. **Fuji, T. and Tonomura, K.,** Oxidation of methanol, formaldehyde and formate by a *Candida* sp., *Agr. Biol. Chem.,* 36, 2297, 1972.
57. **Fuji, T. and Tonomura, K.,** Oxygen dependent oxidation of formaldehyde by cell-free extract of methanol-utilizing yeast, *Candida* sp., *Agr. Biol. Chem.,* 38, 1413, 1974.
58. **Kato, N., Tani, Y., Ogata, K.,** Enzyme system for methanol oxidation in yeasts, *Agr. Biol. Chem.,* 38, 675, 1974.
59. **Davis, J. B., Chase, H. H., and Raymond, R. L.,** *Mycobacterium paraffinicum* sp. a bacterium isolated from soil, *Appl. Microbiol.,* 4, 310, 1956.
60. **Brisbane, P. G. and Ladd, J. N.,** The utilization of methane, ethane, and propane by soil microorganisms, *J. Gen. Appl. Microbiol.,* 14, 447, 1968.
61. **Thomson, A. W., O'Neill, J. G., and Wilkinson, J. F.,** Acetone production by *Methylobacteria, Arch. Mikrobiol.,* 109, 243, 1976.
62. **Lukins, H. B. and Foster, J. W.,** Methyl ketone metabolism in hydrocarbon-utilizing microbacteria, *J. Bacteriol.,* 85, 1074, 1963.
63. **Pabst, G. S. and Brown, L. R.,** The role of isopropyl alcohol in the microbial metabolism of propane, *Develop. Indust. Microbiol.,* 9, 394, 1968.
64. **Vestal, J. R. and Perry, J. J.,** Divergent metabolic pathways for propane and propionate utilization by a soil isolate, *J. Bacteriol.,* 99, 216, 1969.
65. **Vestal, J. R. and Perry, J. J.,** Effect of substrate on the lipids of the hydrocarbon utilizing *Mycobacterium vaccae, Can. J. Microbiol.,* 17, 445, 1971.
66. **Tanaka, K., Ohshima, K., Kimura, K., and Yamamoto, M.,** U.S. patent 3,639,210, 1972.
67. **Blevins, W. T. and Perry, J. J.,** Metabolism of propane, n-propyl-amine, and propionate by hydrocarbon-utilizing bacteria, *J. Bacteriol.,* 112, 513, 1972.
68. **Cerniglia, C. E. and Perry, J. J.,** Crude oil degradation by micoorganisms isolated from the marine environment, *Z. Allg. Mikrobiol.,* 13, 299, 1973.
69. **McLee, A. G., Kormendy, A. C., and Wayman, M.,** Isolation and characterization of n-butane-utilizing microorganisms, *Can. J. Microbiol.,* 18, 1191, 1972.
70. **Kormendy, A. C. and Wayman, M.,** Characteristic cytoplasmic structures in microorganism utilizing n-butane and 1-butanol, *Can. J. Microbiol.,* 20, 225, 1974.
71. **Philips, W. E., Jr. and Perry, J. J.,** Metabolism of n-butane and 2-butanone by *Mycobacterium vaccae, J. Bacteriol.,* 120, 987, 1974.
72. **Foster, J. W.,** Bacterial oxidation of hydrocarbons, in *Oxygenases,* Hayaishi, O. Ed., Academic Press, New York, 1962, 241.
73. **Perry, J. J.,** Microbial cooxidations involving hydrocarbons, *Microbiol. Rev.,* 43, 59, 1979.
74. **Beam, H. W. and Perry, J. J.,** Co-metabolism as a factor in microbial degradation of cycloparaffinic hydrocarbons, *Arch. Mikrobiol.,* 91, 87, 1973.
75. **Beam, H. W. and Perry, J. J.,** Microbial degradation and assimilation of n-alkyl-substituted cycloparaffins, *J. Bacteriol.,* 118, 394, 1974.
76. **Perry, J. J.,** Microbial metabolism of cyclic hydrocarbons and related compounds, *CRC Crit. Rev. Microbiol.,* 387, 1977.
77. **Quayle, J. R.,** Historical prospectives. Proceeding of the Institute of Petroleum Microbiology Group Meeting on Hydrocarbons in Biotechnology P1-2. University of Kent, Canterbury, September 1979.
78. **Abbott, B. J. and Hou, C. T.,** Oxidation of 1-alkenes to 1,2-epoxy-alkanes by *Pseudomonas oleovorans, Appl. Microbiol.,* 26, 86, 1973.
79. **Colby, J. D., Stirling, I., and Dalton, H.,** The soluble methane monooxygenase of *Methylococcus capsulatus* (Bath). Its ability to oxygenate n-alkanes, n-alkenes, ethers, and alicyclic, aromatic and hetero-cyclic compounds, *Biochem. J.,* 165, 395, 1977.
80. **Stirling, D. I. and Dalton, H.,** The fortuitous oxidation and cometabolism of various carbon compounds by whole-cell suspensions of *Methylococcus capsulatus* (Bath), *FEMS Microbiol. Letts.,* 5, 315, 1979.
81. **Higgins, I. J., Hammond, R. C., Sariaslani, F. S., Best, D., Davies, M. M., Tryhorn, S. E., and Taylor, F.,** Biotransformation of hydrocarbons and related compounds by whole organism suspensions of methane-grown *Methylosinus trichosporium* OB3b, *Biochem. Biophys. Res. Commun.,* 89, 671, 1979.
82. **Cetus Company,** *Chemical Week,* Oct. 1979.

83. **Hou, C. T., Patel, R. N., Laskin, A. I., Barnabe, N., and Marczak, I.,** Microbial oxidation of gaseous hydrocarbons: production of methyl ketones from their corresponding secondary alcohols by methane- and methanol-grown microbes, *Appl. Environ. Microbiol.*, 38, 135, 1979.
84. **Hou, C. T., Patel, R. N., Laskin, A. I., Barnabe, N., and Marczak, I.,** Identification and purification of a nicotinamide adenine dinucleotide dependent sec-alcohol dehydrogenase from C_1-utilizing microbes, *FEBS Lett.*, 101, 179, 1979.
85. **Patel, R. N., Hou, C. T., Laskin, A. I., Derelanko, P., and Felix, A.,** Microbial production of methyl ketones. Purification and properties of a secondary alcohol dehydrogenase from yeast, *Eur. J. Biochem.*, 101, 401, 1979.
86. **Machan, L. P. and Heydeman, M. T.,** *Pseudomonas aeruginosa* mutants defective in heptane oxidation, *J. Gen. Microbiol.*, 85, 77, 1974.
87. **Whitworth, D. A., Moo-Young, M., and Viswanathan, T.,** Hydrocarbon fermentations: oxidation mechanism and nonionic-surfactant effects in a culture of *Candida lipolytica, Biotech. Bioeng.*, 15, 649, 1973.
88. **Duppel, W., Lebault, J. M., and Coon, M. J.,** Properties of a yeast cytochrome P_{450}-containing enzyme system which catalyzes the hydroxylation of fatty acids, alkanes, and drugs, *Eur. J. Biochem.*, 35, 583, 1973.
89. **Dalhoff, A. and Rehm, H. J.,** Studies on regulation of tetradecane oxidation in *Pseudomonas aeruginosa*. I. The effect of tetradecane degradation intermediates, especially of C_2 units on tetradecane oxidation, *Eur. J. Appl. Microbiol.*, 3, 193, 1976.
90. **Dalhoff, A. and Rehn, H. J.,** Studies on regulation of tetradecane oxidation in *Pseudomonas aeruginosa*. II. The effect of glucose on tetradecane oxidation, *Eur. J. Appl. Microbiol.*, 3, 203, 1976.
91. **Souw, P., Reiff, J., and Rehn, H. J.,** Oxidation of n-alkanes by citric acid producing *Candida* spp. I. Oxidation and assimilation of n-tetradecane, *Eur. J. Appl. Microbiol.*, 3, 43, 1976.
92. **Hoffman, B. and Rehn, H. J.,** Degradation of long chain n-alkanes by *Murcorales*. I. Development of mucorales on n-alkanes and on some possible oxidation products, *Eur. J. Appl. Microbiol.*, 3, 19, 1976.
93. **Hoffman, B. and Rehn, H. J.,** Degradation of long chain n-alkanes by *Mucorales*. II. Extracellular oxidation products and degradation pathways, *Eur. J. Appl. Microbiol.*, 3, 31, 1976.
94. **Takeda J.,** U.S. Patent 3,616,214, 1968.
95. **Peterson, J. A., Basu, D., and Coon, M. J.,** Enzymatic ω-oxidation. I. Electron carriers in fatty acid and hydrocarbon hydroxylation, *J. Biol. Chem.*, 241, 5162, 1966.
96. **Peterson, J. A., Kusonose, M., Kusonose, E., and Coon, M. J.,** Enzymatic ω-oxidation. II. Function of rubredoxin as the electron carrier in ω-hydroxylation, *J. Biol. Chem.*, 242, 4334, 1967.
97. **Peterson, J. A. and Coon, M. J.,** Enzymatic w-oxidation. III. Purification and properties of rubredoxin, a component of the ω-hydroxylation system of *Pseudomonas oleovorans, J. Biol. Chem.*, 243, 329, 1968.
98. **Nieder, M. and Shapiro, J.,** Physiological function of the *Pseudomonas putida* P_pG_6 (Pseudomonas oleovorans) alkane hydroxylase: monoterminal oxidation of alkanes and fatty acids, *J. Bacteriol.*, 122, 93, 1975.
99. **Grund, A., Shapiro, J., Fennewald, M., Bacha, P., Leahy, J., Markbreiter K., Nieder, M., and Toepfer, M.,** Regulation of alkane oxidation in *Pseudomonas putida, J. Bacteriol.*, 123, 546, 1975.
100. **Benson, S. and Shapiro, J.,** Induction of alkane hydroxylase proteins by unoxidized alkane in *Pseudomonas putida, J. Bacteriol.*, 123, 759, 1975.
101. **Benson, S. and Shapiro, J.,** Plasmid-determined alcohol dehydrogenase activity in alkane-utilizing strains of *Pseudomonas putida, J. Bacteriol.*, 126, 794, 1976.
102. **Boyer, R. F., Lode, E. T., and Coon, M. J.,** Reduction of alkyl hydroperoxides to alcohols: role of rubredoxin, an electron carrier in the bacterial hydroxylation of hydrocarbons, *Biochem. Biophys. Res. Commun.*, 44, 925, 1971.
103. **Shennan, J. L. and Levi, J. D.,** The growth of yeasts on hydrocarbons, *Prog. Indust. Microbiol.*, 13, 1, 1974.
104. **Cardini, G. and Jurtshuk, P.,** Cytochrome P-450 involvement in the oxidation of n-octane by cell-free extracts of *Corynebacterium* sp. strain 7EIC, *J. Biol. Chem.*, 243, 6071, 1968.
105. **Cardini, G. and Jurtshuk, P.,** The enzymatic hydroxylation of n-octane by *Corynebacterium* sp. strain 7EIC, *J. Biol. Chem.*, 245, 2789, 1970.
106. **Gallo, M., Bertrand, J. C., and Azoulay, E.,** Participation du Cytochrome Paso dans L'Oxydation des Alcanes Chez *Candida tropicalis, FEBS Lett.*, 19, 45, 1971.
107. **Gallo, M., Bertrand, J. C., Roche, B., and Azoulay, E.,** Alkane oxidation in *Candida tropicalis, Biochim. Biophys. Acta*, 296, 624, 1973.
108. **Gallo, M., Roche, B., and Azoulay, E.,** Microsomal cytochromes of *Candida tropicalis* grown on alkanes, *Biochim. Biophys. Acta*, 419, 425, 1976.

109. **Bacchin, B., Robertiello, A., and Viglia, A.,** Identification of n-decane oxidation products in *Corynebacterium* cultines by combined gas chromatography-mass spectrometry, *Appl. Microbiol.*, 28, 737, 1974.
110. **Pelz, B. F. and Rehm, H. J.,** Terminale und Subterminale Oxidation von n-Alkanen Durch Schimmelpilze, *Arch. Mikrobiol.*, 92, 153, 1973.
111. **Chen, D. C. T. and Markovetz, A. J.,** Subterminal oxidation of n-tridecane by a cell-free extract of *Penicillium, Zeitschr. Allg. Mikrobiol.*, 14, 525, 1974.
112. **Kachholz, T. and Rehm, H. J.,** Degradation of long chain alkanes by bacilli. I. Development and product formation by bacilli degrading alkanes in the presence of other carbon sources, *Eur. J. Appl. Microbiol.*, 4, 101, 1977.
113. **Ogata, K., Kaneyuki, H., Kato, N., Tani, Y., and Yamada, H.,** Accumulation of decanedioic acid from n-decane by *Torulopsis Candida* No. 99, *J. Ferment. Technol.*, 51, 227, 1973.
114. Mitsui Petrochemical Industries, Japan Patent 9025116, 1972.
115. Kyowa Hakko Kogyo Company, Ltd., Japan Patent 16028708, 1970.
116. **Casida, L. E. and Klein, D. A.,** U.S. Patent 3,625,824, 1970.
117. **Uchio, R. and Shiio, I.,** Tetradecane 1,14-dicarboxylic acid production from n-hexadecane by *Candida cloacae, Agr. Biol. Chem.*, 36, 1389, 1972.
118. **Ladd, J. N.,** The oxidation of hydrocarbons by soil bacteria, *Aust. J. Biol. Sci.*, 9, 92, 1956.
119. **Klausmeier, R. E., Brown, L. R., Benes, E. W., and Strawinski, R. J.,** Propane metabolism by a *Mycobacterium, Bact. Proc.*, 123, 1958.
120. **Klein, D. A., Davis, J. A., and Casida, L. E.,** Oxidation of n-alkenes to ketones by an *Arthrobacter* species, *Antonie van Leeuwenhoek*, 34, 495, 1968.
121. **Formey, F. W. and Markovetz, A. J.,** Subterminal oxidation of aliphatic hydrocarbons, *J. Bacteriol.*, 102, 281, 1970.
122. **Allen, J. E., Formey, F. W., and Markovetz, A. J.,** Microbial subterminal oxidation of alkanes and alk-l-enes, *Lipids*, 6, 448, 1971.
123. **Klug, M. J. and Markovetz, A. J.,** Degradation of hydrocarbons by members of the genus *Candida.* II. Oxidative intermediates from 1-hexadecene and 1-heptadecene by *Candida lipolytica, J. Bacteriol.*, 96, 1115, 1968.
124. **Jones, D. F. and Howe, R.** Microbiological oxidation of long-chain aliphatic compounds. I. Alkanes and alk-1-enes, *J. Chem. Soc. (C)*, 1968, 2801, 1968.
125. **May, S. W. and Abbott, B. J.,** Enzymatic epoxidation. I. Alkene epoxidation by the w-hydroxylation system of *Pseudomonas oleovorans, Biochem. Biophys. Res. Commun.*, 48, 1230, 1972.
126. **May, S. W. and Abbott, B. J.,** Enzymatic epoxidation. II. Comparison between the epoxidation and hydroxylation reactions catalyzed by the w-hydroxylation system of *Pseuddomonas oleovorans, J. Biol. Chem.*, 248, 1725, 1973.
127. **King, D. H. and Perry, J. J.,** The origin of fatty acids in the hydrocarbon-utilizing microorganism *Mycobacterium vaccae, Can. J. Microbiol.*, 21, 85, 1975.
128. **Cerniglia, C. E. and Perry, J. J.,** Effect of substrate on the fatty acid composition of hydrocarbon-utilizing filamentous fungi, *J. Bacteriol.*, 118, 844, 1974.
129. **Yanagawa, S., Tanaka, A., and Fukui, S.,** Fatty acid compositions of *Corynebacterium simplex* grown on 1-alkenes, *Agr. Biol. Chem.*, 36, 2129, 1972.
130. **Cermiglia, C. E., Blevins, W. T., and Perry, J. J.,** Microbial oxidation and assimilation of propylene, *Appl. Environ. Microbiol.*, 32, 764, 1976.
131. **DeBont, J. A. M.,** Oxidation of ethylene by soil bacteria, *Antonie van Leeuwenhoek*, 42, 59, 1976.
132. **DeBont, J. A. M. and Albers, R. A. J. M.,** Microbial metabolism of ethylene, *Antonie van Leeuwenhoek*, 42, 73, 1976.
133. **DeBont, J. A. M., Attwood, M. M., Primrose, S. B., and Harder, W.,** Epoxidation of short chain alkenes in mycobacterium E20: the involvement of a specific mono-oxygenase, *FEMS Microbiol. Lett.*, 6, 183, 1979.
134. **Ohta, H. and Tetsukawa, H.,** Asymmetric epoxidation of long chain terminal olefins by *Corynebacterium equi* IFO 3730, *Agr. Biol. Chem.*, 43, 2099, 1979.
135. **Pirnik, M. P., Atlas, R. M., and Bartha, R.,** Hydrocarbon metabolism by *Brevibacterium erythrogenes:* normal and branched alkanes, *J. Bacteriol.*, 875, 1974.
136. **McKenna, E. J.,** *Degradation of Synthetic Organic Molecules in the Biosphere,* Printing and Publishing Office, National Academy of Science, Washington, D.C., 1972, 73.
137. **Yamada, Y., Motoi, H., Kinoshita, S., Takada, N., and Okada, H.,** Oxidative degradation of squalene by *Arthrobacter* sp., *Appl. Environ. Microbiol.*, 29, 400, 1975.
138. **Yamada, Y., Kusuhada, N., and Okada, H.,** Oxidation of linear terpenes and squalene variants by *Athrobacter* sp., *Appl. Environ. Microbiol.*, 33, 771, 1977.
139. **Higgins, I. J. and Burns, R. G.,** *The Chemistry and Microbiology of Pollution,* Academic Press, London, 1975.

140. **DeKlerk, H. and Van der Linden, A. C.,** Bacterial degradation of cyclohexane. Participation of a co-oxidation reaction, *Antonie van Leeuwenhoek,* 40, 7, 1974.
141. **Ooyama, J. and Foster, J. W.,** Bacterial oxidation of cycloparaffinic hydrocarbons, *Antonie van Leeuwenhoek,* 31, 45, 1965.
142. **Tonge, G. M. and Higgins, I. J.,** Microbial metabolism of alicyclic hydrocarbons. Growth of *Nocardia petroleophila* (NCIB 9438) on methylcyclohexane, *J. Gen. Microbiol.,* 81, 521, 1974.
143. **Stirling, L. A., Watkinson, R. J., and Higgins, I. J.,** Microbial metabolism of alicyclic hydrocarbons: isolation and properties of a cyclohexane-degrading bacterium, *J. Gen. Microbiol.,* 99, 119, 1977.
144. **Norris, D. B. and Trudgill, P. W.,** The metabolism of cyclohexanol by *Nocarbia globerula* CL1, *Biochem. J.,* 121, 363, 1971.
145. **Donoghue, N. A. and Trudgill, P. W.,** The metabolism of cyclohexanol by *Acinetobacter* NC1B.9871, *Eur. J. Biochem.,* 60, 1, 1975.
146. **Donoghue, N. A., Norris, D. B., and Trudgill, P. W.,** Purification and properties of cyclohexanone oxygenase from *Nocardia globerula* CL-1 and an *Acinetobacter* NC1B.9871, *Eur. J. Biochem.,* 63, 175, 1976.
147. **Griffin, M. and Trudgill, P. W.,** Purification and properties of cyclopentanone oxygenase of *Pseudomonas* NC1B.9872, *Eur. J. Biochem.,* 63, 199, 1976.
148. **Ledet, E. J. and Laseter, J. L.,** Alkanes at the air-sea interface from off-shore Louisiana and Florida, *Science,* 186, 261, 1974.
149. **Davis, J. B. and Raymond, R. L.,** Oxidation of alkyl-substituted cyclic hydrocarbons by a *Nocardia* during growth on n-alkanes, *Appl. Microbiol.,* 9, 383, 1961.
150. **Rho, E. M. and Evans, W. C.,** The aerobic metabolism of cyclohexane carboxylic acid by *Acinetobacter anitratum, Biochem. J.,* 148, 11, 1975.
151. **Dutton, R. L. and Evans, W. C.,** The metabolism of aromatic compounds by *Rhodopseudomonas palustris.* A new reductive method of aromatic ring metabolism, *Biochem. J.,* 113, 525, 1969.
152. **Guyer, M. and Hegeman, G.,** Evidence for a reductive pathway for the anaerobic metabolism of benzoate, *J. Bacteriol.,* 99, 906, 1969.
153. **Kaneda, T.,** Enzymatic aromatization of 4-ketocyclohexane carboxylic acid to p-hydroxybenzoic acid, *Biochem. Biophys. Res. Commun.,* 58, 140, 1974.
154. **Blakley, E. R.,** The microbial degradation of cyclohexanecarboxylic acid: a pathway involving aromatization to form p-hydroxybenzoic acid, *Can. J. Microbiol.,* 20, 1297, 1975.
155. **Smith, D. I. and Calley, A. J.,** The microbial degradation of cyclohexane carboxylic acid, *J. Gen. Microbiol.,* 91, 210, 1975.
156. **Prema, B. R. and Bhattacharyya, P. K.,** Microbiological transformation of terpenes. II. Transformations of α-pinene, *Appl. Microbiol.,* 10, 524, 1962.
157. **Shukla, O. P. and Bhattacharyya, P. K.,** Microbiological transformations of terpenes. XI. Pathways of degradation of α- and β-pinenes in a soil pseudomonad (PL-strain), *Indian J. Biochem.,* 5, 92, 1968.
158. **Shukla, O. P., Moholay, M. N., and Bhattacharyya, P. K.,** Microbiological transformations of terpenes. X. Fermentation of α- and β-pinenes by a soil pseudomonad (PL-strain), *Indian J Biochem.,* 5, 79, 1968.
159. **Dhavalikar, R. S. and Bhattacharyya, P. K.,** Microbiological transformations of terpenes. VIII. Fermentation of limonene by a soil pseudomonad, *Indian J. Biochem.,* 3, 144, 1966.
160. **Dhavalikar, R. S., Rangachari, P. N., and Bhattacharyya, P. K.,** Microbiological transformations of terpenes. IX. Pathways of degradation of limonene in a soil pseudomonad, *Indian J. Biochem.,* 3, 158, 1966.
161. **Khanchandani, K. S. and Bhattacharyya, P. K.,** Microbiological transformations of terpenes. XX. Fermentation of camphene by a soil *Pseudomonad* (camphene strain), *Indian J. Biochem.,* 10, 261, 1973.
162. **Khanchandani, K. S. and Bhattacharyya, P. K.,** Microbiological transformations of terpenes. XXI. Growth adaptation studies on the degradation of camphene by the *Pseudomonas* camphene strain, *Indian J. Biochem.,* 11, 110, 1974.
163. **Khanchandani, K. S. and Bhattacharyya, P. K.,** Microbiological transformations of terpenes. XXII. Enzymes involved in the degradation of camphene in the *Pseudomonas* camphene strain, *Indian J. Biochem.,* 11, 179, 1974.
164. **Prema, B. R. and Bhattacharyya, P. K.,** Microbiological transformations of terpenes. III. Transformations of some mono- and sesquiterpenes, *Appl. Microbiol.,* 10, 529, 1962.
165. **Pujar, B. G. and Bhattacharyya, P. K.,** Microbiological transformations of terpenes. XVII. Fermentation of 1-p-methene and 3-p-menthene by a soil pseudomonad (PL-strain), *Indian J. Biochem.,* 10, 164, 1973.
166. **Pujar, B. G. and Bhattacharyya, P. K.,** Microbiological transformations of terpenes. XVIII. Pathway of degradation of 1-p-menthene in a soil pseudomonad (PL-strain), *Indian J. Biochem.,* 10, 170, 1973.

167. **Pujar, B. G., Rangachari, P. N., and Bhattacharyya, P. K.,** Microbiological transformations of terpenes. XIX. Pathway of degradation of 3-p-menthene in a soil pseudomonad (PL-strain), *Indian J. Biochem.*, 10, 173, 1973.
168. **Mukherjee, B. B., Kraidman, G., and Hill, I. D.,** Synthesis of glycols by microbial transformation of some monocyclic terpenes, *Appl. Microbiol.*, 25, 447, 1973.
169. **Tsukamoto, Y., Nonomura, S., and Sakai, H.,** Formation of cis-p-methan-1-ol from pentane by microbial hydroxylation, *Agr. Biol. Chem.*, 39, 617, 1975.
170. **Marr, E. K. and Stone, R. W.,** Bacterial oxidation of benzene, *J. Bacteriol.*, 81, 425, 1961.
171. **Gibson, D. T., Koch, J. R., and Kallio, R. E.,** Oxidative degradation of aromatic hydrocarbons by microorganisms. I. Enzymatic formation of catechol from benzene, *Biochemistry*, 7, 2653, 1968.
172. **Gibson, D. T., Koch, J. R., Schuld, C. L., and Kallio, R. E.,** Oxidative degradation of aromatic hydrocarbons by microorganisms. II. Metabolism of halogenated aromatic hydrocarbons, *Biochemistry*, 7, 3795, 1968.
173. **Gibson, D. T., Cardini, G. E., Maseles, F. C., and Kallio, R. E.,** Incorporation of oxygen-18 into benzene by *Pseudomonas putida*, *Biochemistry*, 9, 1631, 1970.
174. **Hogn, T. and Jaenicke, L.,** Benzene metabolism of *Moraxella* species, *Eur. J. Biochem.*, 30, 369, 1972.
175. **Axcell, B. C. and Geary, P. J.,** The metabolism of benzene by bacteria. Purification and some properties of the enzyme cis-1,2-dihydroxy cyclohexa-3,5-diene (nicotinamide adenine dinucleotide) oxidoreductase (cis-benzene glycol dehydrogenase), *Biochem. J.*, 136, 927, 1973.
176. **Axcell, B. C. and Geary, P. J.,** Purification and some properties of a soluble benzene-oxidizing system from a strain of *Pseudomonas*, *Biochem. J.*, 146, 173, 1975.
177. **Kitagawa, M.,** Studies on the oxidation mechanism of methyl group, *J. Biochem. (Tokyo)*, 43, 553, 1956.
178. **Nozaka, J. and Kusunose, M.,** Metabolism of hydrocarbons in microorganisms. I. Oxidation of p-xylene and toluene by cell-free enzyme preparations of *Pseudomonas aeruginosa*, *Agr. Biol. Chem.*, 32, 1033, 1968.
179. **Worsey, M. J. and Williams, P. A.,** Metabolism of toluene and the xylenes by *Pseudomonas putida* (arvilla) mt-2: evidence for a new function of the TOL plasmid, *J. Bacteriol.*, 124, 7, 1975.
180. **Claus, D. and Walker, N.,** The decomposition of toluene by soil bacteria, *J. Gen. Microbiol.*, 36, 107, 1964.
181. **Nozaka, J. and Kusunose, M.,** Metabolism of hydrocarbons in microorganisms. II. Degradation of toluene by cell free extracts of *Pseudomonas mildenbergii*, *Agr. Biol. Chem.*, 33, 962, 1969.
182. **Gibson, D. T., Hensley, M., Yoshioka, H., and Mabry, R. J.,** Formation of (+)-cis-2,3-dihydroxy-1-methylcyclohexa-4,6-diene from toluene by *Pseudomonas putida*, *Biochemistry*, 9, 1626, 1970.
183. **Yeh, W. K., Gibson, D. T., and Liu, T. N.,** Toluene dioxygenase: a multicomponent enzyme system, *Biochem. Biophys. Res. Commun.*, 78, 401, 1977.
184. **Subramanian, V., Liu, T. N., Yeh, W. K., and Gibson, D. T.,** Toluene dioxygenase: purification of an iron-sulfur protein by affinity chromatography, *Biochem. Biophys. Res. Commun.*, 91, 1131, 1979.
185. **Omori, T., Horiguchi, S., and Yamada, K.,** Studies on the utilization of hydrocarbons by microorganisms. X. Screening of aromatic hydrocarbon-assimilating microorganisms and p-toluic acid formation from p-xylene, *Agr. Biol. Chem.*, 31, 1337, 1967.
186. **Omori, T. and Yamada, K.,** Studies on the utilization of hydrocarbons by microorganisms. XIII. Oxidation of m-xylene and pseudocumene by *Pseudomonas aeruginosa*, *Agr. Biol. Chem.*, 33, 659, 1969.
187. **Omori, T. and Yamada, K.,** Studies on the utilization of hydrocarbons by microorganisms. XVI. Detection of metabolic intermediates of xylene and pseudocumene, *Agr. Biol. Chem.*, 34, 659, 10.
188. **Omori, T. and Yamada, K.,** Studies on the utilization of hydrocarbons by microorganisms. XVII. Metabolism of p-xylene and related compounds, *Agr. Biol. Chem.*, 34, 664, 1970.
189. **Davey, J. F. and Gibson, D. T.,** Bacterial metabolism of para- and meta-xylene: oxidation on a methyl substituent, *J. Bacteriol.*, 119, 923, 1974.
190. **Gibson, D. T., Mahadevan, V., and Davey, J. F.,** Bacterial metabolism of para- and meta-xylene: oxidation of the aromatic ring, *J. Bacteriol.*, 119, 930, 1974.
191. **Omori, T., Jigami, Y., and Minoda, Y.,** Isolation, identification and substrate assimilation specificity of some aromatic hydrocarbon utilizing bacteria, *Agr. Biol. Chem.*, 39, 1775, 1975.
192. **Jigami, Y., Omori, T., and Minoda, Y.,** The degradation of isopropylbenzene and isobutylbenzene by *Pseudomonas* sp., *Agr. Biol. Chem.*, 39, 1781, 1975.
193. **Jamison, V. W., Raymond, R. L., and Hudson, J. O.,** in *Proc. 3rd Int. Biodegradation Symp.*, Sharpley, J. M. and Kaplan, A. M., Eds., Applied Science, London, 1976, 187.
194. **Baggi, G., Catelani, D., Galli, E., and Treccani, V.,** The microbial degradation of phenylalkanes: 2-phenylbutane, 3-phenylpentane, 3-penyldodecane, and 4-phenylheptane, *Biochem. J.*, 126, 1091, 1972.

195. **Jigami, Y., Omori, T., and Minoda, Y.,** A new ring fission product suggestive of an unknown reductive step in the degradation of n-butylbenzene by *Pseudomonas, Agr. Biol. Chem.,* 38, 1757, 1974.
196. **Sielicki, M., Focht, D. D., and Martin, J. P.,** Microbial transformations of styrene and [^{14}C] styrene in soil and enrichment cultures, *Appl. Environ. Microbiol.,* 35, 124, 1978.
197. **Shirai, K. and Hisatsuka, K.,** Production of phenethyl alcohol from styrene by *Pseudomonas* 305-STR -1-4, *Agr. Biol. Chem.,* 43, 1399, 1979.
198. **Jigami, Y., Kawasaki, Y., Omori, T., and Minoda, Y.,** Coexistence of different pathways in the metabolism of n-propylbenzene by *Pseudomonas* sp., *Appl. Environ. Microbiol.,* 38, 783, 1979.
199. **Webley, D. M., Duff, R. B., and Farmer, V. C.,** Evidence for oxidation in the metabolism of saturated aliphatic hydrocarbons by soil species on *Norcardia, Nature (London),* 178, 1467, 1956.
200. **Douros, J. D. and Frankenfeld, J. W.,** Oxidation of alkylbenzenes by a strain of *Micrococcus cerificans* growing on n-paraffins, *Appl. Microbiol.,* 18, 532, 1968.
201. **Jigami, Y., Omori, T., and Minoda, Y.,** Screening of n-alkylbenzenes assimilating yeasts and identification of oxidation products from n-alkylbenzenes, *Agr. Biol. Chem.,* 38, 401, 1974.
202. **McKenna, E. J. and Coon, M. J.,** Enzymatic α-oxidation. IV. Purification and properties of the α-hydroxylase of *Pseudomonas oleovorans, J. Biol. Chem.,* 246, 3882, 1971.
203. **Jurtshuk, P. and Cardini, G. E.,** The mechanism of hydrocarbon oxidation by a *Corynebacterium* species, *Crit. Rev. Microbiol.,* 1, 239, 1971.
204. **Van Ravenswaay Claasen, J. C. and van der Linden, A. C.,** Substrate specificity of the paraffin hydroxylase of *Pseudomonas aeruginosa, Antonie van Leeuwenhoek,* 37, 339, 1971.
205. **Gibson, D. T., Gschwendt, B., Yeh, W. K., and Kobal, V. M.,** Initial reactions in the oxidation of ethylbenzene by *Pseudomonas putida, Biochemistry,* 12, 1520, 1973.
206. **Hopper, D. J.,** The hydroxylation of p-cresol and its conversion to p-hydroxybenzaldehyde in *Pseudomonas putida, Biochem. Biophys. Res. Commun.,* 69, 462, 1976.
207. **Nozaki, M.,** *Molecular Mechanism of Oxygen Activation,* Hayaishi, Ed., Academic Press, New York, 1974.
208. **Dagley, S.,** Catabolism of aromatic compounds by microorganisms, Adv. Microbiol. Physiol., 6, 1, 1971.
209. **Chapman, P. J.,** *Degradation of Synthetic Organic Molecules in the Biosphere,* Natl. Acad. Sci., Washington, D.C., 1972, 17.
210. **Ferris, J. P., Fasco, M. J., Stylianopoulou, F. L. Jerina, D. M., Daly, J. W., and Jeffrey, A. M.,** Mono-oxygenase activity in *Cunninghamella bainieri:* evidence for a fungal system similar to liver microsomes, *Arch. Biochem. Biophys.,* 156, 97, 1973.
211. **Partch, J.,** *Chemical Economic Handbook,* Stanford Research Institute, Menlo Park, Calif., 1976.
212. **Davis, J. I. and Evans, W. C.,** Oxidative metabolism of naphthalene by soil pseudomonards. The ring-fission mechanism, *Biochem. J.,* 91, 251, 1964.
213. **Barnsley, E. A.,** Naphthalene metabolism by pseudomonads: The oxidation of 1,2-dihydroxynaphthalene to 2-hydroxychromene-2-carboxylic acid and the formation of 2-hydroxybenzalpyruvate, *Biochem. Biophys. Res. Commun.,* 72, 1116, 1976.
214. **Ferris, J. P., MacDonald, L. H., Patrie, M. A., and Martin, M. A.,** Aryl hydrocarbon hydroxylase activity in the fungus *Cunninghamella bainieri:* evidence for the presence of cytochrome P-450, *Arch. Biochem. Biophys.,* 175, 443, 1976.
215. **Smith, R. V. and Rosazza, J. P.,** Microbial models of mammaliam metabolism. Aromatic hydrocarbon, *Arch. Biochem. Biophys.,* 161, 551, 1974.
216. **Wegner, E. H.,** U.S. Patent No. 3,796,630, 1974.
217. **Cerniglia, C. E. and Gibson, D. T.,** Metabolism of naphthalene by cell extracts of *Cunninghamella elegans, Arch. Biochem. Biophys.,* 186, 121, 1978.
218. **Cerniglia, C. E. and Gibson, D. T.,** Metabolism of naphthalene by *Cunninghamella elegans, Appl. Environ. Microbiol.,* 34, 363, 1977.
219. **Cerniglia, C. E., Gibson, D. T., and Van Baalen, C.,** Algal oxidation of aromatic hydrocarbons: formation of 1-naphthol from naphthalene by *Agmenellum quadruplicatum,* strain PR-6, *Biochem. Biophys. Res. Commun.,* 88, 50, 1979.
220. **Cerniglia, C. E., Van Baalen, C., and Gibson, D. T.,** Metabolism of naphthalene by *Cyanobacterium oscillatoria* sp., strain JCM, *J. Gen. Microbiol.,* 116, 484, 1980.
221. **Cerniglia, C. E., Gibson, D. T., and Van Baalen, C.,** Algal oxidation of naphthalene, *J. Gen. Microbiol.,* 116, 494, 1980.
222. **Tsfasman, I. M., Starovoitov, I. I., Zyakun, A. M., and Skrybin, G. K.,** Initial reaction of enzymatic oxidation of 1,8-dimethyl naphthalene, *Biokhimiya,* 41, 864, 1976.
223. **Kodama, K., Umehada, K., Shimizu, K., Nakatani, S., Minoda, Y., and Yamada, K.,** Identification of microbial products from dibenzothiophene and its proposed oxidation pathway, *Agri. Biol. Chem.,* 37, 45, 1973.
224. **Hou, C. T. and Laskin, A. I.,** Microbial conversion of dibenzothiophene, *Dev. Ind. Microbiol.,* 17, 351, 1976.

225. **Evans, W. C., Fernley, H. N., and Griffiths, E.,** Oxidative metabolism of phenanthrene and anthracene by soil pseudomonads. The ring-fission mechanism, *Biochem. J.*, 95, 819, 1965.
226. **Kiyohara, H., Nagao, K., and Nomi, R.,** Degradation of phenanthrene through o-phthalate by an *Aeromonas sp.*, *Agr. Biol. Chem.*, 40, 1075, 1976.
227. **Jerina, D. M., Selander, H., Yagi, H., Wells, M. C., Davey, J. F., Mahadevan, V., and Gibson, D. T.,** Dihydrodiols from anthrancene and phenanthrene, *J. Am. Chem. Soc.*, 98, 5988, 1976.
228. **Gibson, D. T., Mahadevan, V., Jerina, D. M., Yagai, H., and Yeh, H. J. C.,** Oxidation of the carcinogens benzo[a]pyrene and benzo[a]anthracene to dihydrodiols by a bacterium, *Science*, 189, 295, 1975.
229. **Gibson, D. T.,** *Fate and Effects of Petroleum Hydrocarbons in Marine Ecosystems and Organisms*, Wolfe, D. A., Ed., Pergamon Press, Elmsford, New York, 1977, 36.
230. **Cerniglia, C. E. and Gibson, D. T.,** Oxidation of benzo(a)pyrene by the filamentous fungus *Cunninghamella elegans*, *J. Biol. Chem.*, 254, 12175, 1979.
231. **Kiyohara, H.,** Production of a red benzaanthracene from kerosene by *Pseudomonas* sp. 57K5, *Agr. Biol. Chem.*, 43, 1407, 1979.
232. **Catelani, D., Sorlini, C., and Trecanni, V.,** The metabolism of biphenyl by *Pseudomonas putida*, *Experimentia*, 27, 1173, 1971.
233. **Catelani, D., Colombi, A., Sorlini, C., and Treccani, V.,** Metabolism of biphenyl. 2-Hydroxy-6-oxo-6-phenyl hexa-2,4-dienoate: the meta-cleavage product from 2,3-dihydroxybiphenyl by *Pseudomonas putida*, *Biochem. J.*, 134, 1063, 1973.
234. **Catelani, D. and Columbi, A.,** Metabolism of biphenyl. Structure and physicochemical properties of 2-hydroxy-6-phenylhexa-2,4 dienoic acid, the meta-cleavage product from 2,3-dihydroxybiphenyl by *Pseudomonas putida*, *Biochem. J.*, 143, 431, 1974.
235. **Lunt, D. and Evans, W. C.,** The microbial metabolism of biphenyl, *Biochem. J.*, 118, 54, 1970.
236. **Dodge, R. H., Cerniglia, C. E., and Gibson, D. T.,** Fungal metabolism of biphenyl, *Biochem. J.*, 178, 223, 1979.
237. **Raymond, R. L. and Jamison, V. W.,** Biochemical activities of *Nocardia*, *Adv. Appl. Microbiol.*, 14, 93, 1971.
238. **Gibson, D. T.,** The microbial oxidation of aromatic hydrocarbons, *Crit. Rev. Microbiol.*, 1, 199, 1971.
239. **Chakrabarty, A. M.,** Genetic basis of the biodegradation of salicylate in *Pseudomonas*, *J. Bacteriol.*, 112, 815, 1972.
240. **Rheinwald, J. G., Chakrabarty, A. M., and Gunsalus, I. C.,** A transmissible plasmid controlling camphor oxidation in *Pseudomonas putida*, *Proc. Natl. Acad. Sci. U.S.A.*, 70, 885, 1973.
241. **Chakrabarty, A. M., Chou, G., Gunsalus, I. C.,** Genetic regulation of octane dissimilation plasmid in *Pseudomonas*, *Proc. Natl. Acad. Sci. U.S.A.*, 70, 1137, 1973.
242. **Grund, A., Shapiro, J., Fennewald, M., Bacha, P., Leahy, J., Markbreiter, K., Neider, M., and Toepfer, M.,** Regulation of alkane oxidation in *Pseudomonas putida*, *J. Bacteriol.*, 123, 546, 1975.
243. **Nieder, M. and Shapiro, J.,** Physiological function of the *Pseudomonas putida* P_pG_o *(Pseudomonas oleovorans)* alkane hydroxylase: monoterminal oxidation of alkanes and fatty acids, *J. Bacteriol.*, 122, 93, 1975.
244. **Benson, S., Oppici, M., Shapiro, J., and Fennewald, M.,** Regulation of membrane peptides by the *Pseudomonas* plasmid alk ruglon, *J. Bacteriol.*, 140, 754, 1979.
245. **Dunn, N. W. and Gunsalus, I. C.,** Transmissible plasmid coding early enzymes of naphthalene oxidation in *Pseudomonas putida*, *J. Bacteriol.*, 114, 974, 1973.
246. **Williams, P. A. and Worsey, M. J.,** Ubiquity of plasmids in coding for toluene and xylene metabolism in soil bacteria: evidence for the existence of new TOL plasmids, *J. Bacteriol.*, 125, 818, 1976.
247. **Friello, D. A., Malroie, J. R., and Chakrabarty, A. M.,** *Proc. 3rd. Intl. Biodegradation Symp.*, Sharpley, J. M. and Kaplan, A. M., Eds., Applied Science Publishers Ltd., London, 1976, 205.
248. **Friello, D. A., Malroie, J. R. Gibson, D. T., Rogers, J. E. and Chakrabarty, A. M.,** XYL, a non-conjugative xylene-degradative plasmid in *Pseudomonas PXY*, *J. Bacteriol.*, 127, 1217, 1976.
249. **Williams, P. A.,** Microbial genetics relating to hydrocarbons degradation, in *Developments in Biodegradation of Hydrocarbons I*, Watkinson, R. J., Ed., Applied Science Publishers, 1978, 135.

Chapter 5

MICROBIAL TRANSFORMATION OF PROSTAGLANDINS

George P. Peruzzotti

TABLE OF CONTENTS

I. INTRODUCTION AND SCOPE

One of the most striking and significant developments in synthetic organic chemistry is the use of microbial systems to effect chemical transformations. Microorganisms possess adaptive enzyme systems which can, under suitable conditions, be raised to high levels to produce products in substantial quantities. Microbial enzymes are known which will catalyze most of the common transformations of organic syntheses such as oxidations, reductions, hydrolysis, and condensations. These enzymes are specific with respect to the reaction they catalyze and the position of attack on a substrate. Moreover, many enzymes are stereospecific as well as regiospecific and effect their changes under mild physiological conditions. Because of their high specificity, such systems in many cases have distinct advantages over usual organic reactions.

Prostaglandins are a family of naturally occurring hormone-like substances derived biosynthetically from C-20 polyunsaturated fatty acids containing as many as three or four double bonds. They are potent vasoactive substances that play a key role in regulating cellular metabolism and function in minute quantities. Structurally they are derivatives of prostanoic acid and are characterized by a five-membered carbocyclic ring with two side chains extending from adjacent carbon atoms on the ring. Nominally they are classed into types and series according to the functionalities in the cyclopentane ring and the degree of unsaturation in the side chains (See Figure 1).

The precursors of prostaglandins are C-20 essential fatty acids which exist in tissues as phospholipids. Bergström et al.[1,2] and Van Dorp et al.[3,4] demonstrated that 8,11,14-eicosatrienoic acid was the biological precursor of PGE_1, 5,8,11,14 eicosatetraenoic acid (arachidonic acid) was the precursor of PGE_2 and 5,8,11,14,17 eicosapentaenoic acid leads to PGE_3.

These essential fatty acids are made available for prostaglandin synthesizing enzymes by phospholipase A. The enzymatic synthesis of prostaglandins of the "2" series is depicted in Figure 2.[62]

Prostaglandins are very rapidly metabolized by degradative enzymes.[5] In man, the initial metabolic step is oxidation of the 15 alcohol to a ketone function followed by reduction of the 13,(14) double bond, beta-oxidation of the acid side chain and ω-1 and ω-2 hydroxylation of the alkyl side chain. Hamberg and Israelsson have described liver enzymes which convert the 9 keto prostaglandins to the 9α and 9β hydroxy isomers.[6] Similar enzymes have also been found in microorganisms.

Prostaglandins have been intensively studied because of their profound effects on physiological processes. They are local hormones because they act in the immediate area in which they are produced, as opposed to circulating hormones which operate at distant targets. Present in the body in nanogram quantities, these substances exhibit a strikingly diverse range of biological effects, such as lowering or raising blood pressure, constricting bronchi, stimulating or relaxing uterine smooth muscle, inhibiting gastric secretion, mediating inflammation, promoting sodium excretion, and inducing labor. Although prostaglandins may have a wide range of effects, they do show strict structure specificity, e.g., PGE_1 inhibits, whereas PGE_2 enhances platelet aggregation.[7]

Since the prostaglandin molecule can have as many as five asymmetric centers, total chemical synthesis represents a difficult task. Conventional chemical synthesis requires, at some stage, an optical resolution in order to obtain the correct optical antipode. Chemical resolving agents theoretically yield 50% of product with the desired configuration.[8] Microbial transformations, which are essentially enzyme reactions, are very selective to substrate used and product formed and well suited to obtain a particular optical isomer in high yield. The use of microbial systems and enzymes as chemical reagents has been amply reviewed.[9-13]

Various aspects of prostaglandin transformation by microorganisms have appeared

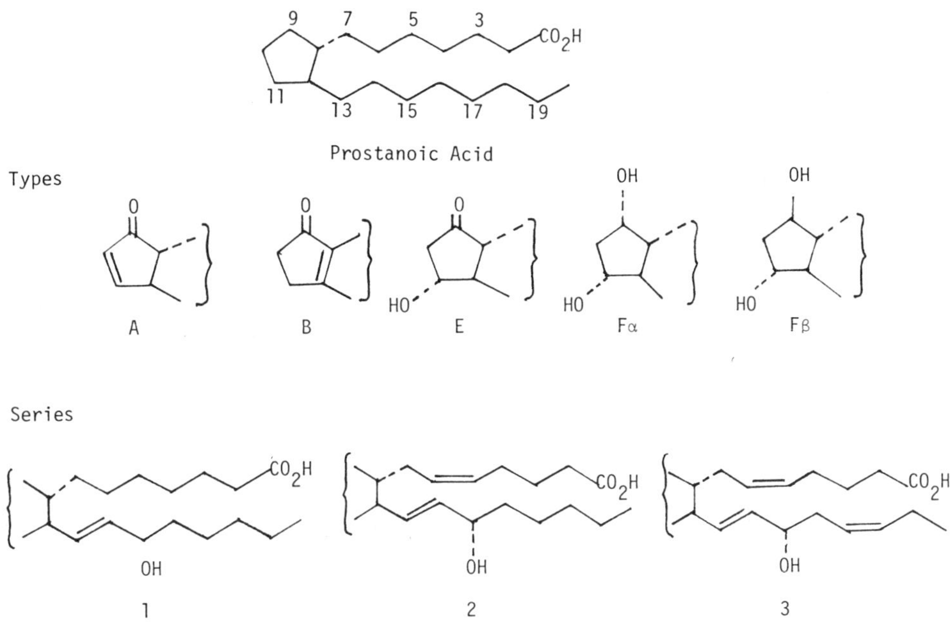

Prostaglandin Nomenclature

in earlier reviews.[10,13,14] The purpose of this review is to summarize the application of microbial systems to prostaglandin transformations, draw attention to their usefulness in organic synthesis, and show similarities between microbial and mammalian metabolism of prostaglandins. The reviewer has taken the liberty to artificially divide this subject matter into two broad categories, namely; microbial transformation of (1) prostaglandin synthons, and (2) prostaglandins and prostaglandin analogs.

II. PROSTAGLANDIN SYNTHONS

Introduction of chiral centers into the prostaglandin molecule has been the subject of intensive studies by several groups. Since natural prostaglandins have as many as four or five chiral centers, the development of synthetic methods to attain this goal is crucial. One synthetic approach toward the total synthesis of prostaglandins was the development by Corey's group of an optically active lactone (1) as a key synthon with four chiral centers.[15]

A microbial approach to (1) was accomplished by reduction of racemic (2) by *Saccharomyces drosophylarum* to yield (−) endo and (+) exo alcohols (3) and (4) which are readily separable.[16] These alcohols are useful precursors in the synthesis of prostanoids since one of them can be chemically transformed into the Corey lactone. (1).

A bioorganic asymmetric total synthesis of prostaglandins was designed by Sih et al.[17,18] where the required asymmetry was introduced microbiologically. The strategy employed by Sih was to view the construction of the prostanoic acid skeleton by conjugate addition of C-8 vinylcuprates to α-alkylated cyclopentenones.

The formation of (5) was accomplished by microbial reduction of readily available 2-(6-carbomethoxyhexyl)-cyclopentane-1,3,4-trione (7) to the 4(R) alcohol 2-(6-carbomethoxyhexyl)-4(R)-hydroxycyclopentane-1,3-dione[8] with *Dipodascus uninucleatus,* whereas *Mucor rammanianus* gave the 4(S) alcohol (9). The E_2 analog of (7) was also reduced by *Dipodascus uninucleatus* to give the E_2 analog of (8).[19]

Chemical conversion of (8) to the enol mesitylene sulfonate, reduction of the C-3 carbonyl function followed by acidic allylic rearrangement, afforded (5) in very good yields.

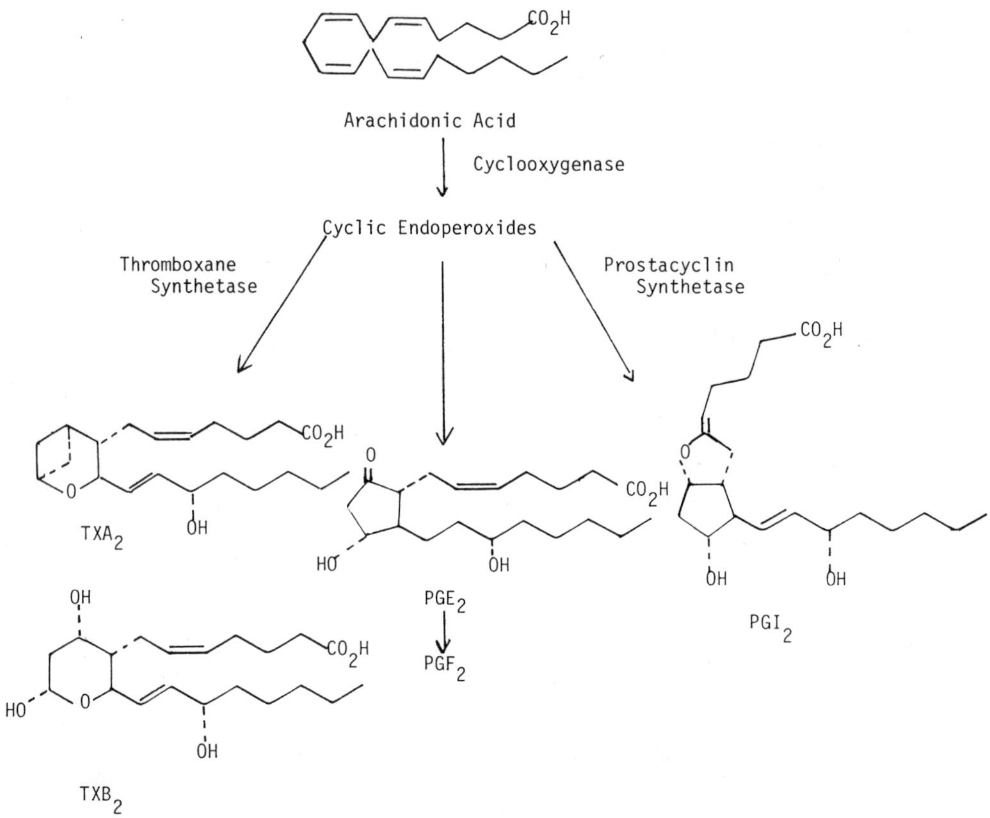

FIGURE 2

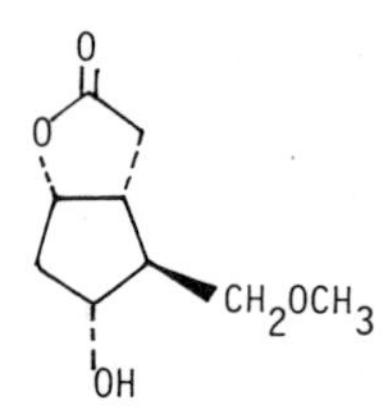

(1)

FIGURE 3

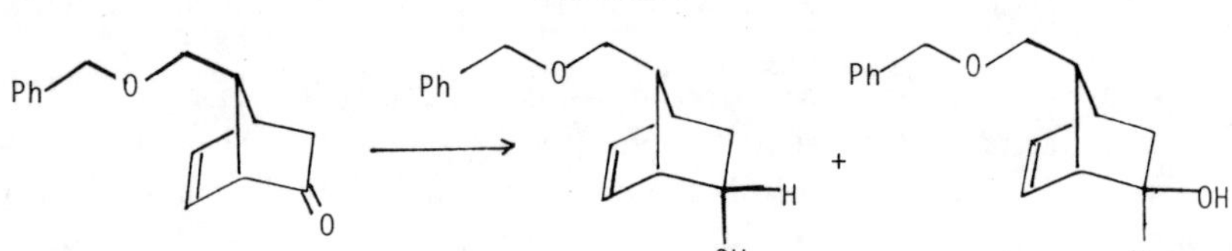

(2) (3) (4)

FIGURE 4

Attention was then focused on the asymmetric synthesis of the octenyl synthon (6). The stereoselective reduction of *trans*-1-iodo-octen-3-one (10) was catalyzed by *Penicillium decumbens* to produce the desired 3(S) alcohol (11), whereas *Aspergillus ustus* gave the 3(R) enantiomer (12).

Having achieved the asymmetric synthesis of (11) and (5), there remained the chemical assemblage of (11) to (5) by conjugate addition of the organocuprate derived from (11) to (5) after protection of the hydroxyl groups to yield PGE$_1$ methyl ester. Ester hydrolysis of the methyl ester with bakers' yeast or *Rhizopus oryzae* afforded PGE$_1$

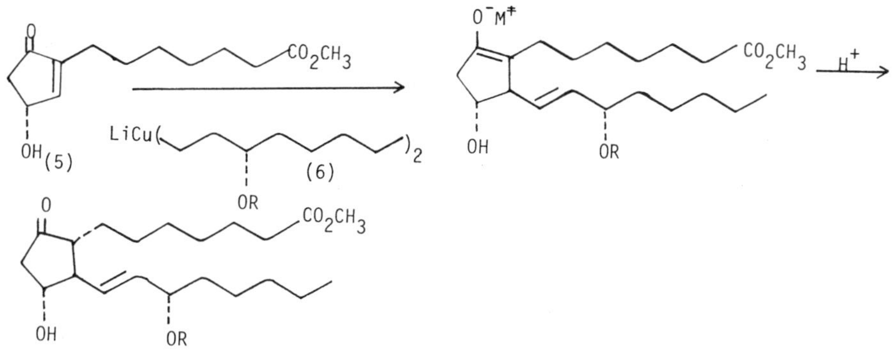

FIGURE 5

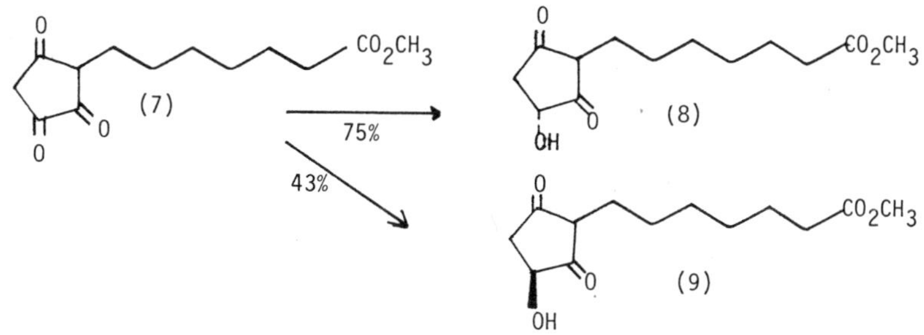

FIGURE 6

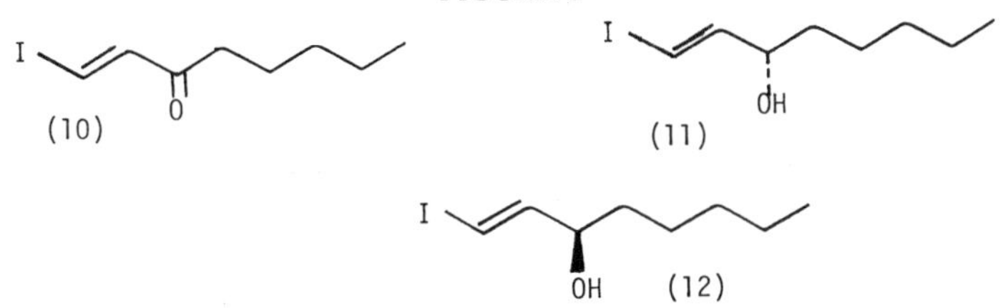

FIGURE 7

acid. Thus the total synthesis of (−) PGE_1 and (−) PGE_2 was accomplished without the use of chemical resolving agents.

Attempts at the chemical resolution of the phthalate half ester of 1-ocytn-4-ol (13), an intermediate in the synthesis of 16-hydroxy prostaglandins, were ineffective, whereas the microbial resolution[20] of the benzoate of (14) by resting cells of *Rhizopus nigricans* yielded the optically active alcohol (15).

Additional support for the use of microorganisms in prostaglandin synthesis was furnished recently by Sih et al.[21] Exposure of mesaconic acid (16) to mesaconase of *Clostridium tetanomorpham* yielded (S)-citramalate (17) which is chemically converted to (2S)-2-methyl-2-hydroxy-1-hexanol (18), a chiral synthon for the synthesis of 15-deoxy-16(S)-hydroxy-16-methylprostaglandins.

Hydroxylation of cinerone (19) to cinerolone (20) by *Aspergillus niger* ATCC 9142[22] served as a model for allylic hydroxylation of the prostaglandin synthon (21) to optically active 2-(4-hydroxy-1-oxocyclopent-2-ene) hepatonoic acid.[23]

Optically active prostaglandin intermediates,[24] 4(R) and 4(S) hydroxycyclopent-2-en-1-ones(23), were synthesized from 3(R), 5(R)-diacetoxycyclopent-1-ene(25), 3(R)-acetoxy-5(R)-hydroxycyclopent-1-ene(26), and 3(S), 5(S)-dihdroxycyclopent-1-ene(27), obtained by microbiological hydrolysis[25] of a 1:1 mixture of *cis* and *trans*-3,5-diacetoxycyclopent-1-ene(24).

Hydrolysis of nonchiral *cis*-3,5-diacetoxycyclopent-1-ene (28) by *Bacillus subtilis var. Niger* gave the chiral monoacetate (29).[26] This was converted chemically to the lactone (30) which Partridge et al.[27] have shown leads to the asymmetric synthesis of $PGF_{2\alpha}$.

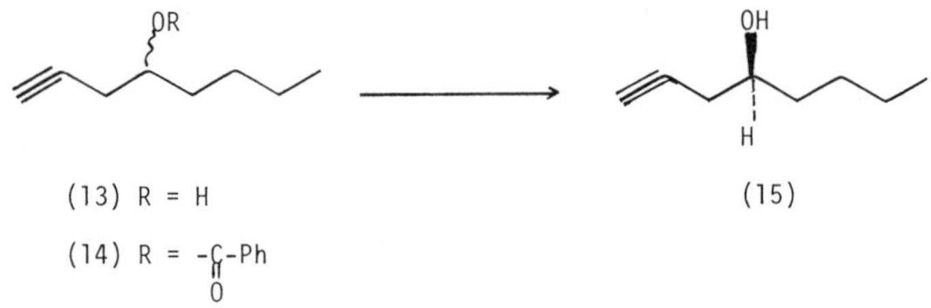

FIGURE 8

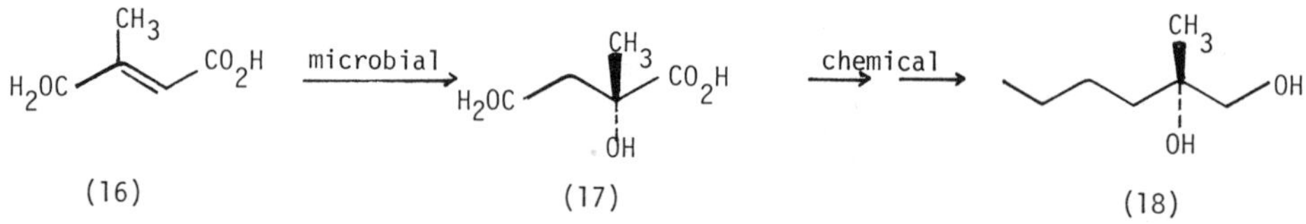

FIGURE 9

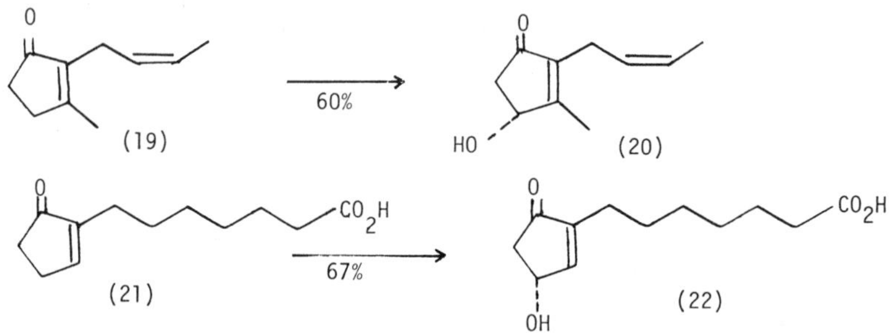

FIGURE 10

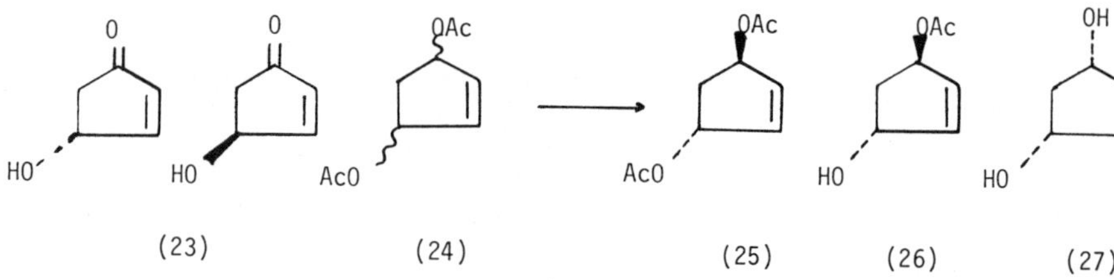

FIGURE 11

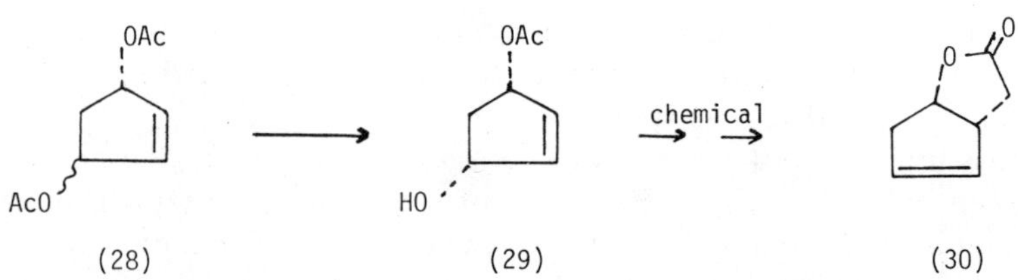

FIGURE 12

Microbial reduction[28] of racemic bicyclo[3.2.0]hept-2-en-6-one (31) by actively fermenting bakers' yeast gave the diastereomeric alcohols (32) and (33) which, after separation, were used to form the enantiomeric bromohydrins (34) and (35).

The enantiomer (34) could be converted to $PGF_{2\alpha}$ via a tricyclic ketone, and (35) yielded $PGF_{2\alpha}$ via the epoxide.

A recent report documents the transformation of cyclopentanedione derivatives (36-41) to chiral lactones (42) and (43) using *Dipodascus albidus*.[29] These chiral lactones are used as intermediates in the synthesis of prostaglandins.

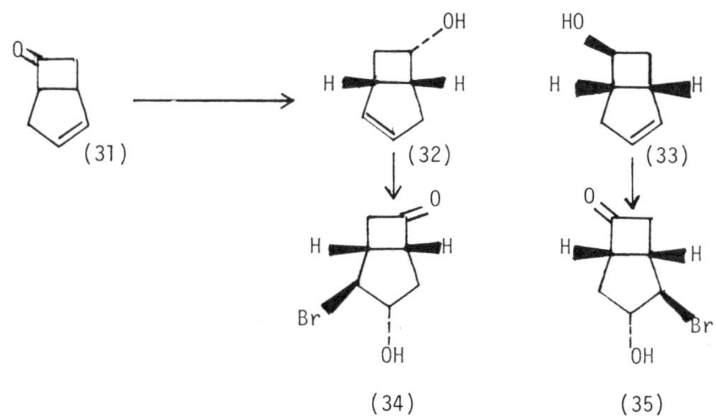

FIGURE 13

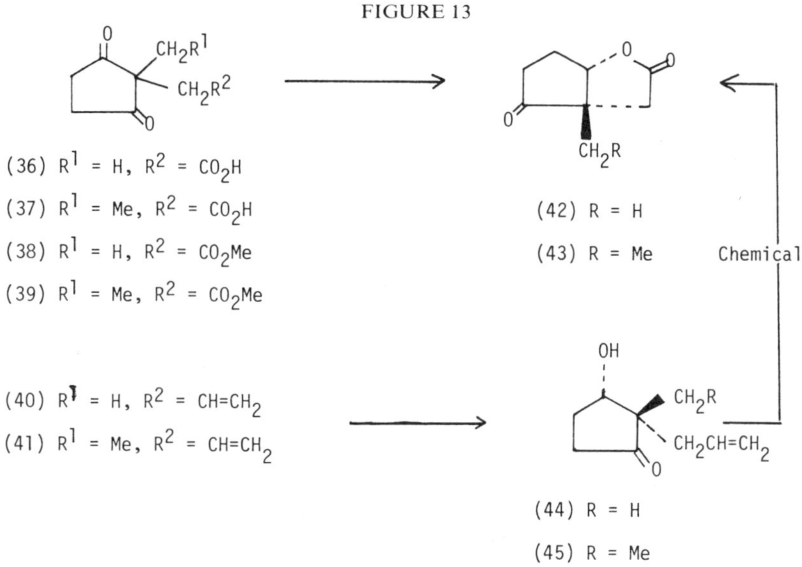

FIGURE 14

III. PROSTAGLANDINS AND PROSTAGLANDIN ANALOGS

A. Hydroxylation

Fungi in general are noted for hydroxylation at nonactivated carbons in a variety of substrates such as alkaloids, steroids, and hydrocarbons.[11] Prostaglandins possess an alkyl side chain which lends itself to hydroxylation. Studies with human seminal plasma have shown that a substantial part of the prostaglandins are in the form of C-19 hydroxylated derivatives.[30] Several reports document that microorganisms also possess this capability to hydroxylate prostaglandins at positions C-18, 19, and 20. One of the earliest reports of C-18, C-19 hydroxylation using microorganisms was reported by Sih et al.[31] Exposure of arachidonic acid (46) to *Ophiobolus graminis* afforded two products, 18ξ arachadonic acid (47) and 19ξ arachidonic acid (48). The resulting products were cyclized by exposure to bull seminal vesicle microsomes (BSVM) to yield the oxygentated prostadienoic acids (49) and (50).

Incubation of 15(S)acetoxy PGA_2 methyl ester (51) with *Streptomyces ruber* (NRRLB-1268) yielded the C-19 hydroxylated derivatives (52) and (53).[32] In addition to hydroxylation at C-19, the 10(11) double bond of the substrate was reduced and hydrolysis of the esters were observed.

Several other species of *Streptomyces* were useful, but *S. ruber* was more efficient. Marx and Doodewaard,[33] using a variety of microorganisms, describe numerous examples of hydroxylations of prostaglandin E_1, E_2, B_2, A_2, $F_{2\alpha}$, and several prostaglan-

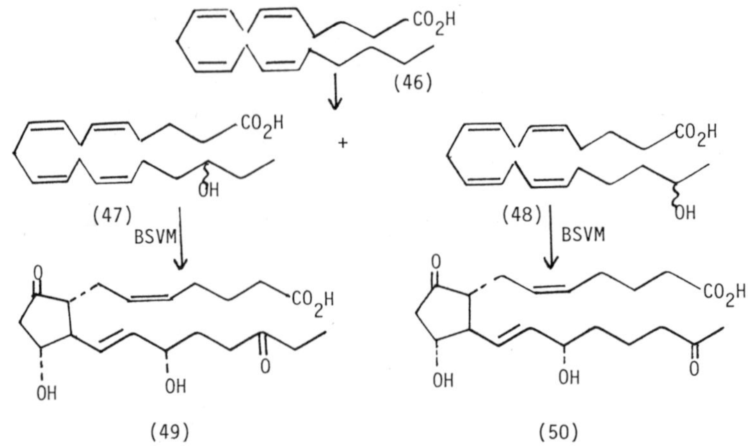

FIGURE 15

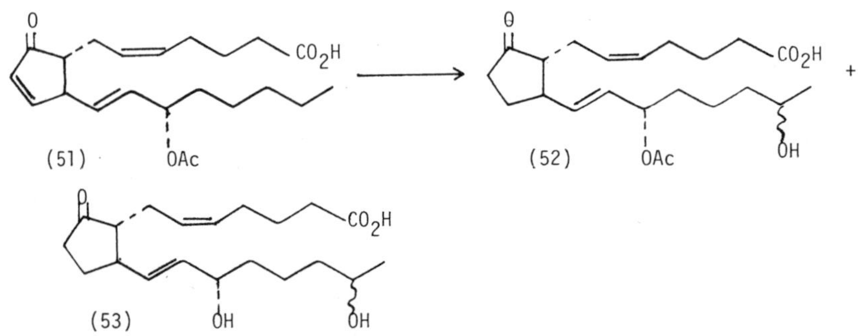

FIGURE 16

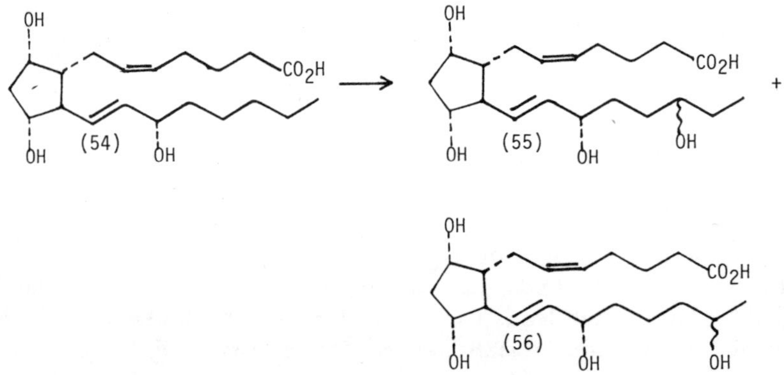

FIGURE 17

din analogs at C-18, C-19, and C-20. In some cases trace amounts of C-17 hydroxylation was observed. Thus *Streptomyces* (CBS 188.74) hydroxylated PGE_1 at C-18 while $PGF_{2\alpha}$ and PGE_2 each gave two products: one hydroxylated at C-18 and the other at C-19. Hydroxylation at C-20 was accomplished using *Aspergillus niger* (ATCC 9142) and PGB_2 as the substrate. The configuration of the hydroxyl group was not determined in any of the examples cited. Hydroxylation at C-18 accompanied by reduction of the 10(11) double bond was reported with *Cunninghamella blakesleeana.* (ATCC 9245) and PGA_2 as substrate.[34] Recent reports substantiate the similarity between microbial and mammalian transformation of prostaglandins.[13,14] Hydroxylation of $PGF_{2\alpha}$ and PGE_2 at C-18 and C-19 by *Streptomyces* sp. (UC5761) was reported by Sebek et al.[35] When $PGF_{2\alpha}$ (54) was exposed to *Streptomyces* sp., 18-hydroxy $PGF_{2\alpha}$ (55) and 19-hydroxy $PGF_{2\alpha}$ (56) were isolated in a combined yield of 20 to 24%. Incu-

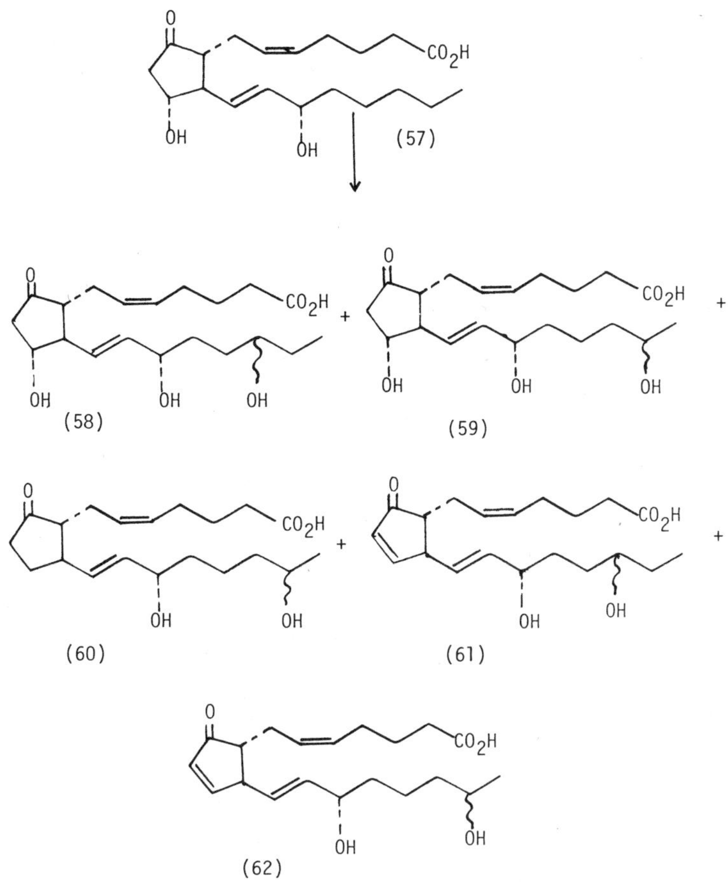

FIGURE 18

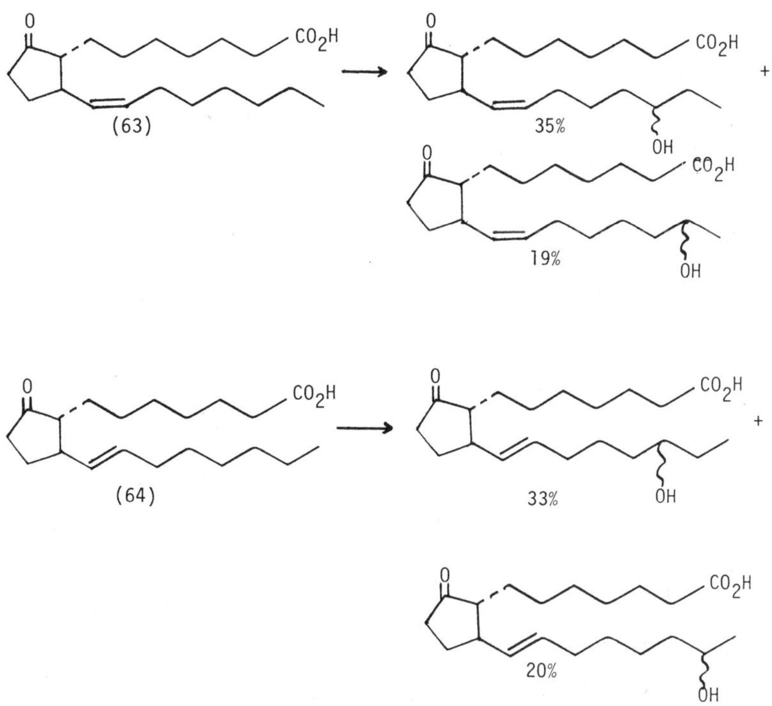

FIGURE 19

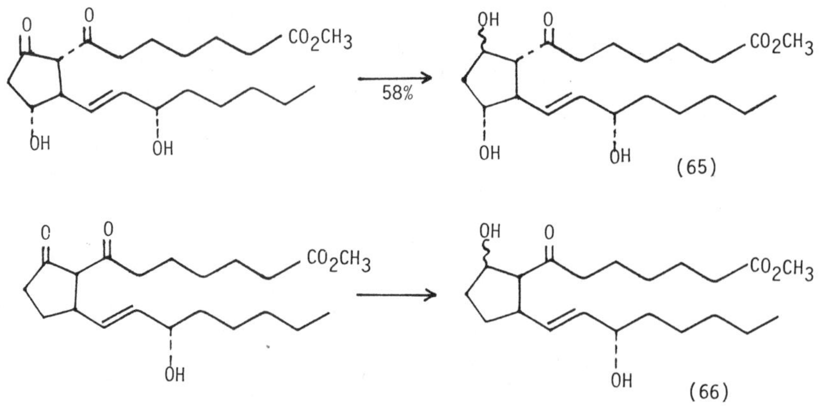

FIGURE 20

bation of PGE_2 (57) with the same organism gave several conversion products (58 to 62) in 40% yield.

The formation of PGA_2 derivatives (61,62) was suggested to be a nonenzymatic dehydration of the substrate during incubation or the isolation procedure. Simple prostanoic acid derivatives such as racemic 9-oxo-13(*cis*) and 9-oxo-13(*trans*) prostenoic acids (63,64) was reported by Lanzilotta et al. to have been hydroxylated at C-18 and C-19 by *Microascus trigonosporus* (1199).[36] Structure elucidation using proton and carbon-13 magnetic resonance indicate some degree of selectivity in hydroxylating the racemic 13 (*trans*) starting material.

B. Reductions

1. Carbonyl Function

a. 9 Keto Reduction

Chemical reduction of the 9 keto function of prostaglandin leads to isomeric mixtures, whereas stereoselective reduction may be obtained using microbial systems. Schneider and Murray employed actively fermenting bakers' yeast to stereoselectively reduce the 9 keto function of PGE_1 and PGE_2 to $PGF_{1\alpha}$ and $PGF_{2\alpha}$, respectively.[37] When racemic PGE_1 and PGE_2 methyl esters were used as substrates, the yeast reduced both enantiomers of the racemates. The stereoselective reduction of racemic 15-deoxy PGE_1 to give 15-deoxy $PGF_{1\alpha}$ and ent-15-deoxy $PGF_{1\beta}$ by *Saccharomyces cerevisiae* was obtained by Sih.[38] The preparation of 7-oxo and 11-deoxy-7-oxo PGF_1 (65,66) from their respective PGE_1 precursors by bakers' yeast is described by Miura et al.[39]

b. 15 Keto Reduction

Microbial stereoselective reduction and optical resolution of racemic $\Delta^{8,12}$-15-dehydro PGE_1 (67) was achieved by Miyano et al.[40] and Marsheck and Miyano.[41] *Flavobacterium* sp. (NRRL-B-5641) reduced the 15 ketone of (67) to produce a *cis* racemic mixture (68,69). Optically active 11-epi $\Delta^{8(12)}$ PGE_1 (70) was prepared in good yield by *Pseudomonas* sp. (NRRL B 3875) when exposed to (67), whereas *Flavobacterium* sp. (NRRL B-3874) or *Arthrobacter* sp. (NRRL-B 3873) gave optically active 15 epi $\Delta^{8(12)}$ PGE_1 (71).

The use of basidiomycetes to stereoselectively reduce the 15 keto group of prostaglandin-like compounds has been extensively detailed by Moore.[42] In all cases cited the 15α alcohol was produced (Table 1). Similarly, Kieslich documents the reduction of the prostaglandin intermediate (72) by *Kloeckera jensenii* (ATCC 20110) to the 15α alcohol in analytical yields of 70% and 60% as the isolated product (73).[43]

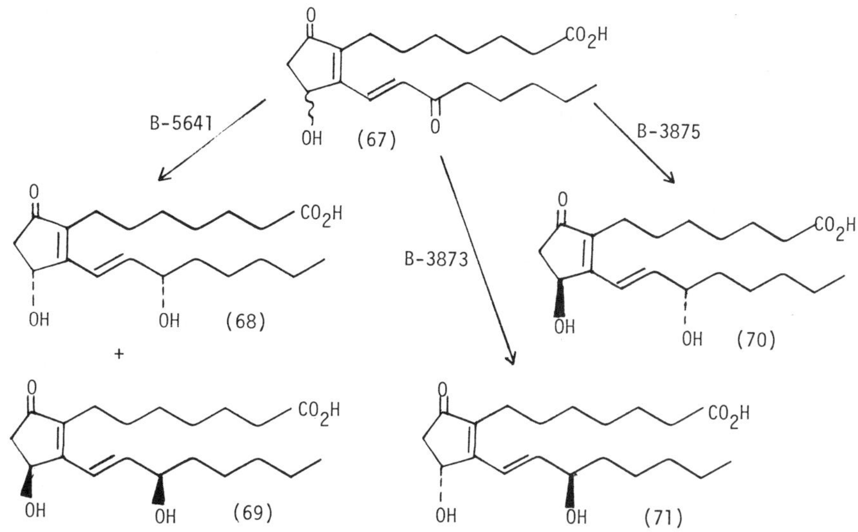

FIGURE 21

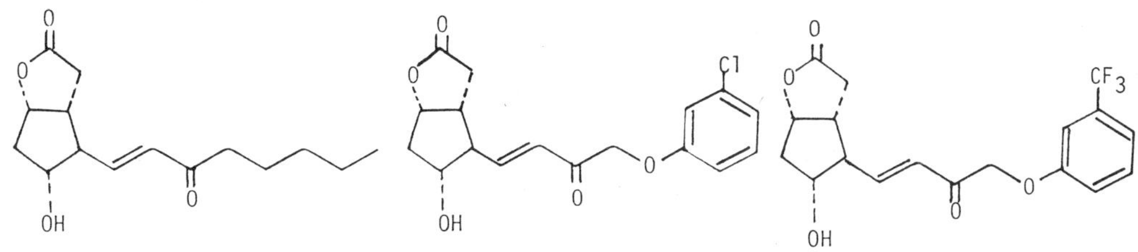

Trechispora brinkmannii
Brettanomyces bruxellensis
Lentinellus montanus
Armillariella mella

Trechispora brinkmannii
T. nivea-cremea
T. raduloides
Creatosporella gordanichii
Armillariella mella
Sistotrema brinkmannii
Heliomyces roseus
Brettanomyces bruxellensis
Sistotrema oblongisporum
Femsjonia luteo-alba

Trechispora brinkmannii
Brettanomyces bruxellensis
Femsjonia luteo-alba
Sistotrema brinkmannii

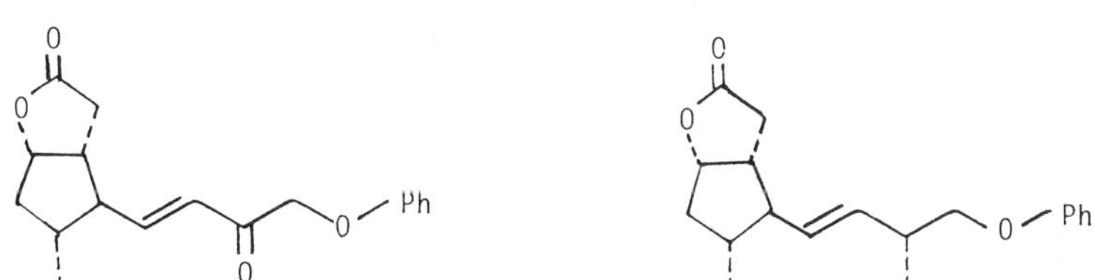

(72) (73)

FIGURE 22

2. Double Bond Reduction

Reduction of double bonds at 10(11) and 13(14) in prostaglandins has been achieved by a wide variety of microorganisms as a side reaction in examples previously cited. The primary focus of the section is to report those papers which cite double bond reduction on the major product. Thus, PGA_2 (74) and 15-epi PGA_2 (75) were reduced by *Streptomyces griseus* to 11 deoxy PGE_2 (76) and 11-deoxy-15-epi PGE_2 (77).[44] Organisms of the genera *Streptomyces, Pseudomonas,* and *Corynebacterium* may also be used. Similarly, *Corynespora cassiicola* was used to reduce the 10(11) double bond of 15-epi PGA_2 methyl ester, acetate. Hydrolysis of the esters was observed.[45] Studies of the microbiological transformation of PGA_2 (74) by workers at Searle[46,47] report

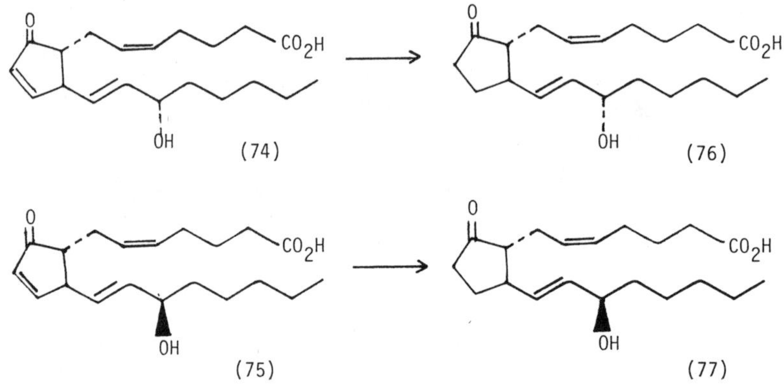

FIGURE 23

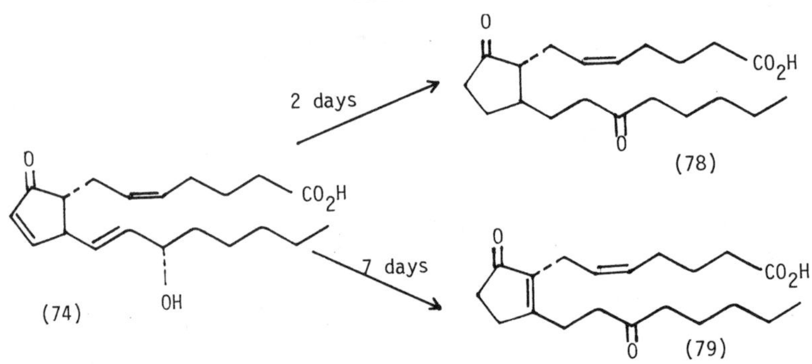

FIGURE 24

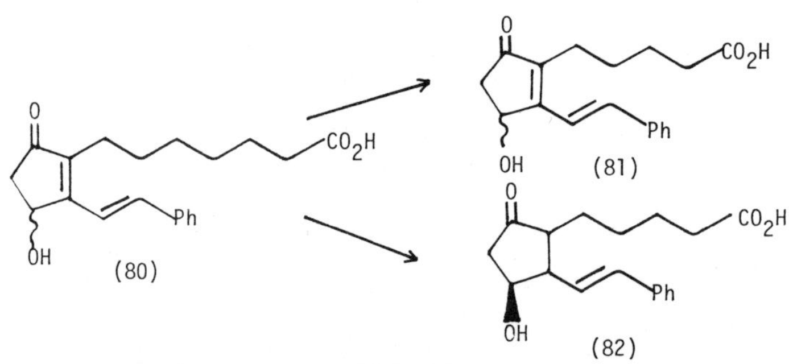

FIGURE 25

the reduction of the 10(11) double bond to yield 11-deoxy PGE_2 (76) by *Bacillus polymyxa* (ATCC 12321), *Gliocladium* sp. (NRRL 5605) and *Cephalosporium* sp. (NRRL 5499). However, *Dactylium denroides* (NRRL 2575) reduced the 10(11) and 13(14) double bond and oxidized the 15α alcohol to the 15 ketone (78). This is analogous to the primary mode of inactivation of prostaglandins in mammalian systems. After 7 days exposure, unsaturation was introduced at 8(12) to form 15-keto-13(14) dihydro PGB_2 (79).

Additionally, *Saccharomyces cerevisiae* is capable of reducing the 13(14) double bond of 15-keto $PGF_{2\alpha}$ in 46% yield.[35]

C. Beta-Oxidation

Studies of the metabolism of prostaglandins in animal and mammalian systems have identified di and tetranor derivatives as a consequence of beta-oxidation.[48] Microbial beta-oxidation of prostaglandins has also been reported. Hsu et al.[49] reported that the racemic styryl prostaglandin analog (80) was beta-oxidized to the dinor derivative (81)

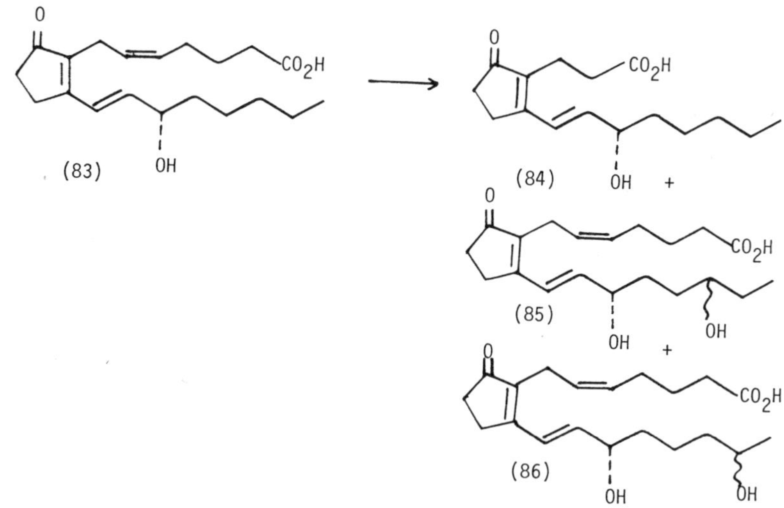

FIGURE 26

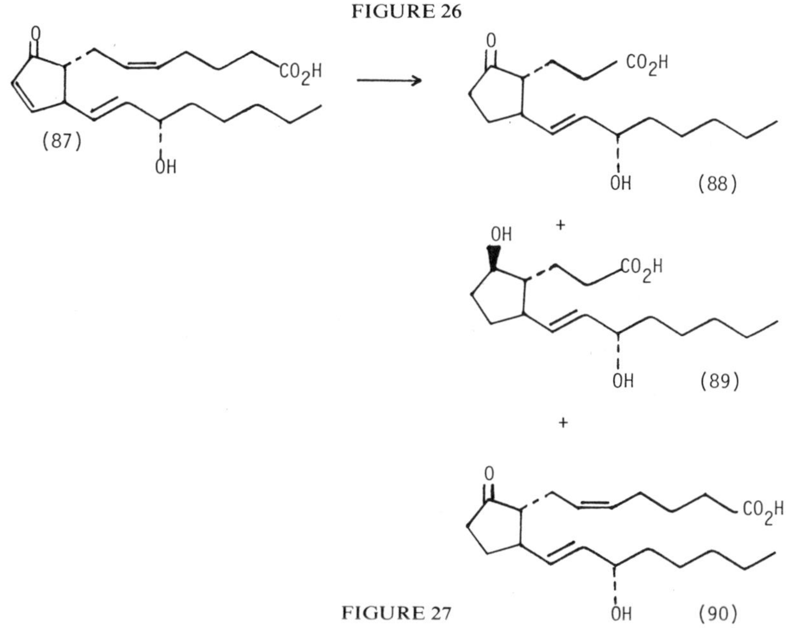

FIGURE 27

by *Penicillium* sp. (M8904), whereas *Mortierella* sp. (M8872) beta-oxidized (80) stereospecifically to optically active (82).

Natural prostaglandins were used to elucidate the scope of this microbial beta-oxidation by *Penicillium* sp. (M8904). This organism beta-oxidizes PGB_2 (83) and PGA_2 (87) to the tetranoic derivatives (84,88,89). Several other transformed products were identified, namely hydroxylation at C-18 or C-19 (85,86), reduction of the 10(11) double bond of PGA_2 (88,89,90), and 9 keto reduction (89).

Studies evaluating PGA_2 solutions for stability revealed an inadvertent contaminant *Aureobasidium pullulans* beta-oxidized PGA_2 to several tetranor metabolites of which three were characterized (91,92,93).[50] *A. pullulans* metabolizes 15-epi PGA_2 to similar products. An effort to supply adequate amounts of mammalian metabolites via microbial transformation was documented by Upjohn workers using *Mycobacterium rhodochrous* (UC6176).[51,52] Thus, beta-oxidation of 15(S)-15-methyl-PGE_2 (94) yielded the dinor derivative (95) in 41% yield. Similarly, 6-keto $PGF_{1\alpha}$ (96), PGI_1 (98) and

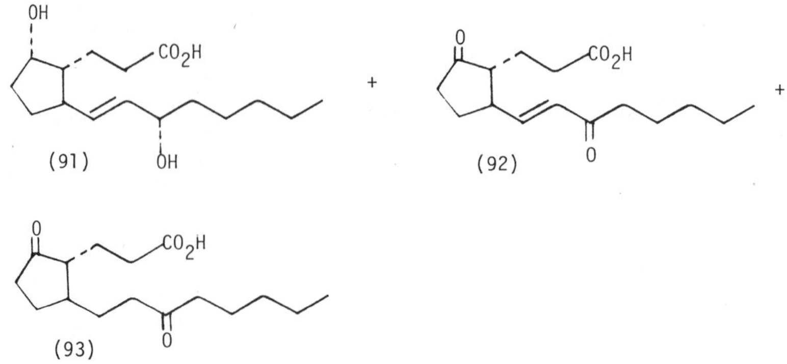

FIGURE 28

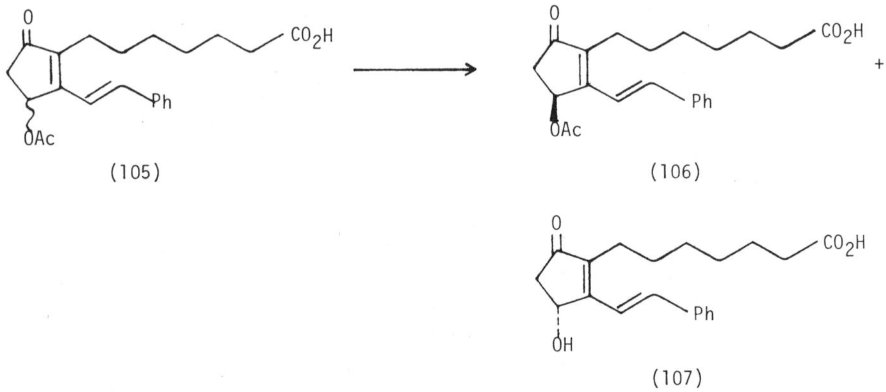

FIGURE 29

thromboxane B_2 (100) were transformed to the corresponding dinor derivatives (97,99,101). Extended incubation with 9-deoxo-16,16-dimethyl-9-methylene-PGE_2 (102) allowed further beta-oxidation and gave (103,104) in 21 and 18% yields respectively. The 2,3 dinor analog (101) was found to be the major urinary metabolite in the monkey, guinea pig, and man.

D. Hydrolysis

Hydrolytic cleavage of prostaglandin esters has been observed in both microorganisms and mammalian systems. Microbial esterases used as a synthetic tool provide an extremely mild method for removal of protecting groups and may show selectivity, thereby increasing their use as resolving agents. Actively fermenting bakers' yeast hydrolyzed the methyl esters of racemic PGE and PGE_2 in addition to the stereoselective reduction and resolution of these substrates.[37] The hydrolytic ability of *Rhizopus oryzae* or bakers' yeast, was employed as a final preparative step for the total synthesis of several prostaglandin acids.[19,53-56] The microbiological hydrolysis of a prostaglandin diester was described by Leeming and Greenspan.[57] *Corynespora cassicola* (IMI 56007) was used to hydrolyze the 15(R) acetoxy and methyl ester of 15(R) acetoxy PGA_2 methyl ester. The stereoselective hydrolysis of a racemic styryl prostaglandin intermediate (105) by *Saccharomyces* sp. (1375-143) gave mostly starting material enriched in the 3(S) enantiomorph (106) and a small amount of the resolved product (107).[58] Hydrolysis of the propionyl and isobutyryl racemic esters was also observed; however, the isobutyryl ester was hydrolyzed less efficiently. *Saccharomyces* sp. (NRRL Y 7342) also hydrolyzed (105) to (103).[59] Cell-free extracts of *Cladosporium resinae* (ATCC 274) have been used by Axen and Murray to hydrolyze PGE methyl esters.[60]

E. Cyclization

A comprehensive patent by workers at Upjohn describes an oxidative cyclization procedure of suitable polyunsaturated fatty acids to various prostaglandins.[61] These transformations are carried out by five classes of fungi, namely *Phycomyces, Ascomyces, Basidiomyces, Deuteromyces* and *Schizomyces*. For example, arachidonic acid was transformed by members of the above classes to PGE_1, PGE_2, $PGF_{1\alpha}$, and $PGF_{2\alpha}$. No yields are given in the description.

IV. SUMMARY

Continued interest in the use of microorganisms as sources of organic chemical reagents to produce chemical transformation of pharmaceutically important organic compounds is amply demonstrated by microbial transformations of prostaglandins. Because of their ease of handling and the mild physiological conditions under which they operate, microbial systems have been employed with great success to introduce chirality or as chiral resolving agents, in the synthesis of prostaglandin synthons and key prostaglandin intermediates not readily available by chemical methods. The stereospecific total synthesis of prostaglandins has also been realized by the judicious application of microbial technology in conjunction with established chemical procedures. Equally noteworthy is the important documentation that microorganisms mimic mammalian metabolism and have been used to supply prostaglandin and thromboxane metabolites. From a practical standpoint, the microbiological transformation of prostaglandins represents a valuable adjunct to the chemical synthesis of natural prostaglandins and their analogs.

ADDENDUM

Since the preparation of this manuscript, an additional product of beta-oxidation of 6-keto PGF_1 to 13,14-dihydro-6,15-diketo-2,3-dinor-PGF_1 by Mycobacterium rodochrous (UC6176) has been reported.[63] Additionally, this organism possesses enzyme systems for dehydrogenation of the 15-hydroxy function and reduction of the 13,14-double bond of prostaglandins which have previously been found in mammalian tissues.

REFERENCES

1. **Bergström, S., Danielsson, H., Klenberg, D., and Samuelsson, B.,** *J. Biol. Chem,* 239 PC, 4006, 1964.
2. **Bergström, S., Danielsson, H., and Sammuelsson, B.,** *Biochem. Biophys. Acta,* 90, 207, 1964.
3. **van Dorp, D. A., Beerthius, R. K., Nugteren, D. H., and Vonkeman, H.,** *Biochem. Biophys. Acta,* 203, 839, 1964.
4. **van Dorp, D. A., Beerthius, R. K., Nugsteren, D. H., and Vonkeman, H.,** *Nature (London),* 203, 839, 1964.
5. **Sammuelsson, B., Granström, E., Green, K., Hamberg, M., and Hammarström, S.,** *Ann. Rev. Biochem.,* 44, 669, 1975.
6. **Hamberg, M. and Isarelsson, U.,** *J. Biol Chem.,* 245, 5107, 1970.
7. **Andersen, N. H. and Ramwell, P. W.,** *Arch. Intern. Med.,* 133, 30, 1974.
8. **Sih, C. J.** *Am. J. Pharma. Ed.,* 41, 432, 1977.
9. **Sih, C. J. and Rosazza, J. P.,** *Application of Biochemical Systems in Organic Chemistry* (Part 1), J. B. Jones, C. J. Sih, and D. Perlman, Eds., J. Wiley & Sons, New York 1976, 69.
10. **Sih, C. J. and Abushanab, E.,** *Ann. Repts. Med. Chem.,* 12, 298, 1977.

11. **Fonken, G. S. and Johnson, R. A.,** *Chemical Oxidations with Microorganisms,* Marcel Dekker, New York, 1972.
12. **Beukers, R., Marx, A. F., and Zuidweg, M. H. J.,** *Drug Design,* Vol. 3, Ariëns, C. J., Ed., Academic Press, New York, 1972, 1.
13. **Rosazza, J. P. and Smith, R. V.,** *Adv. Appl. Microbiol.,* 25, 169, 1979.
14. **Sebek, O. K. and Kieslich, K.,** *Ann. Repts. Ferm. Proc.,* 1, 269, 1977.
15. **Corey, E. J., Weinshenker, N. M., Schaaf, T. K., and Huber, W.,** *J. Amer. Chem. Soc.,* 91, 5675, 1969.
16. **Ambrus, G., Szentirmai, A., Mehesfalvi, C., Kovacs, G., Szantay, C., and Novak, L.,** Hung. Teljes 8066 (1974): Chem. Abstr. 81:49337 b (1974).
17. **Sih, C. J., Salomon, R. G., Price, P., Sood, R., and Peruzzotti, G.,** *J. Am. Chem. Soc.,* 97, 857, 1975.
18. **Sih, C. J., Heather, J. B., Sood, R., Price, P., Peruzzotti, G., Hsu, L. F., and Lee, S. S.,** *J. Am. Chem. Soc.,* 97, 865, 1975.
19. **Heather, J. B., Sood, R., Price, P., Peruzzotti, G., Lee, S. S., Hsu Lee, L. F., and Sih, C. J.,** *Tetrahedron Lett.,* 2313, 1973.
20. **McGahren, W. J., Sax, K. J., Kunstmann, P., and Ellestad, G. A.,** *J. Org. Chem.,* 42, 1659, 1977.
21. **Fujimoto, Y., Yadev, J. S., and Sih, C. J.,** *Tetrahedron Lett.,* 2481, 1980.
22. **Tabenkin, B., LeMahieu, R. H., Berger, J., and Kierstad, R. W.,** *Appl. Microbiol.,* 17, 714, 1969.
23. **Kurozumi, S., Toru, T., and Ishimoto, S.,** *Tetrahedron Lett.,* 4959, 1973.
24. **Tanaka, T., Kurozumi, S., Toru, T., Miura, S., Kobayashi, M., and Ishimoto, S.,** *Tetrahedron Lett.,* 32, 1713, 1976.
25. **Miura, S., Kurozumi, S., Toru, T., Tanaka, T., Kobayashi, M., Matsubara, S., and Ishimoto, S.,** *Tetrahedron Lett.,* 32, 1893, 1976.
26. **Takano, S., Tanigawa, K., and Ogasawara, K.,** *J. Chem. Soc. Chem. Comm.,* 189, 1976.
27. **Partridge, J. J., Chadha, N. K., and Uskokovic, M. R.,** *J. Am. Chem. Soc.,* 95, 7171, 1973.
28. **Newton, R. F., Paton, J., Reynolds, D. P., and Young, S.,** *J. Chem. Soc. Chem. Comm.,* 908, 1979.
29. **Truckenbrodt, G., Schwarz, S., Schick, H., Hess, L., Meyer, M., Weber, G., Porwol, K., Eberhardt, U., and Henkel, H., Ger. (East) 134, 953, (1977);** *Chem. Abstr.,* 91, 173388 q, 1979.
30. **Hamberg, M. and Sammuelsson, B.,** *J. Biol. Chem.,* 241, 257, 1966.
31. **Sih, C. J., Ambrus, G., Foss, P., and Lai, C. J.,** *J. Am. Chem. Soc.,* 91, 3685, 1969.
32. **Marsheck, W. and Miyano, M.,** U.S. Patent 3,878,046, 1975.
33. **Marx, A. F. and Doodewaard, J.,** Brit. Patent 1,501,864, 1978.
34. **Hsu, C. F., Jui, J., and Mizuba, S. S.,** U.S. Patent 3,856,852, 1974.
35. **Sebek, O. K. Lincoln, F. H. and Schneider, W. P.,** *5th Int. Ferment. Cong., West Berlin,* (Abstr.), 17.05, 1976.
36. **Lanzilotta, R. P., Bradley, D. G., McDonald, K. M., and Tökés, L.,** *Appl. Environ. Microbiol.,* 32, 726, 1976.
37. **Schneider, W. P., and Murray, H. C.,** *J. Org. Chem.,* 38, 397, 1973.
38. **Sih, C. J.,** unpublished in Reference 9, p. 74.
39. **Miura, S., Kurozumi, S., Toru, T., Tanaka, T., Kobayoshi, M., and Ishimoto, S.,** Japan Kokai 77,128,354, 1977; *Chem. Abstr.,* 88, 119440a.
40. **Miyano, M., Dorn, C. R., Colton, F. B., and Marsheck, W. J.,** *J. Chem. Soc. Chem. Comm.,* 425, 1971.
41. **Marsheck, W. J. and Miyano, M.,** *Biochem. Biophys. Acta,* 348, 263, 1974.
42. **Moore, R. H.,** U.S. Patent 3,880,712, 1975.
43. **Kieslich, K., Raduechel, B., Skuballa, W., Vorbrueggen, H., and Dahl, H.,** Ger. Offen. 2,853,637, 1980; *Chem. Abstr.,* 93,130606h, .
44. **Greenspan, G. and Leeming, M. R. G.,** U.S. Patent 3,930,952, 1976; *Chem. Abstr.,* 85, 31622p, .
45. **Greenspan, G. and Leeming, M. R. G.,** U.S. Patent 3,928,134, 1975; *Chem. Abstr.,* 84,10384w, .
46. **Hsu, C. F., Jiu, J., and Mizuba, S.,** U.S. Patent 3,788,947, 1974.
47. **Jiu, J., Hsu, C. F., and Mizuba, S.,** *Dev. Ind. Microbiol.,* 15, 345, 1974.
48. **Sammuelsson, B., Granström, E., Green, K., and Hamburg, M.,** *Ann. N.Y. Acad. Sci.,* 180, 138, 1971.
49. **Hsu, C. F. J., Jui, J., and Mizuba, S.,** *Dev. Ind. Microbiol.,* 18, 487, 1977.
50. **Leavey, E. M. K., Andersen, N. H., and Bissonette, P.,** *Prostglandins,* 17, 19, 1979.
51. **Sebek, O. K., Lincoln, F. H., Bundy, G. L., Sih, J. C., Nash, S. A., Jackson, R. W., and Nelson, N. A.,** *6th Int. Ferment. Symp., London, Ontario* (Abstr.), F-20.5(P), 1980.
52. **Nelson, N. A., Jackson, R. W., and Sebek, O. K.,** *Prostaglandins,* 16, 85, 1978.
53. **Sih, C. J., Salomon, R. G., Price, P., Peruzzotti, G., and Sood, R.,** *J. Chem. Soc. Chem. Comm.,* 240, 1972.

54. **Sih, C. J., Price, P., Sood, R., Salomon, R. G., Peruzzotti, G., and Casey, M.,** *J. Am. Chem. Soc.*, 94, 3643, 1972.
55. **Sih, C. J., Salomon, R. G., Price, P., Sood, R., and Peruzzotti, G.,** *Tetrahedron Lett.*, 2435, 1972.
56. **Sih, C. J., Heather, J. B., Peruzzotti, G., Price, P., Sood, R., Hsu Lee, L. F.,** *J. Am. Chem. Soc.*, 95, 1676, 1973.
57. **Leeming, M. R. G. and Greenspan, G.,** U.S. Patent 3,726,765, 1973.
58. **Marsheck, W. J. and Miyano, M.,** *Biochem. Biophys. Acta*, 316, 363, 1973.
59. **Colton, F. B., Marsheck, W. J., Miyano, M.,** Ger. Offen. 2,318,594, 1973; *Chem. Abstr.*, 80:259432Z.
60. **Axen, U. F. and Murray, H. C.,** Ger. Offen. 1,937,678, 1976; *Chem. Abstr.*, 72,109830h, .
61. **Beal, P. F., Fonken, G. S., and Pike, J. E.,** U.S. Patent 3,290,226, 1966.
62. **Moncada, S. and Vane, J. R.,** *N. Engl. J. Med.*, 300, 1142, 1979.
63. **Sun, F. F., Taylor, B. M., Lincoln, F. H., and Sebek, O. K.,** *Prostaglandins*, 20, 729, 1980.

INDEX

A

B

C

D

E

F

G

H

I

K

L

M

R

S

T

V

W

X

Y

Z